Critical Studies of Education

Volume 14

We live in an era where forms of education designed to win the consent of students, teachers, and the public to the inevitability of a neo-liberal, market-driven process of globalization are being developed around the world. In these hegemonic modes of pedagogy questions about issues of race, class, gender, sexuality, colonialism, religion, and other social dynamics are simply not asked. Indeed, questions about the social spaces where pedagogy takes place—in schools, media, corporate think tanks, etc.—are not raised. When these concerns are connected with queries such as the following, we begin to move into a serious study of pedagogy: What knowledge is of the most worth? Whose knowledge should be taught? What role does power play in the educational process? How are new media re-shaping as well as perpetuating what happens in education? How is knowledge produced in a corporatized politics of knowledge? What socio-political role do schools play in the twenty-first century? What is an educated person? What is intelligence? How important are socio-cultural contextual factors in shaping what goes on in education? Can schools be more than a tool of the new American (and its Western allies') twenty-first century empire? How do we educate well-informed, creative teachers? What roles should schools play in a democratic society? What roles should media play in a democratic society? Is education in a democratic society different than in a totalitarian society? What is a democratic society? How is globalization affecting education? How does our view of mind shape the way we think of education? How does affect and emotion shape the educational process? What are the forces that shape educational purpose in different societies? These, of course, are just a few examples of the questions that need to be asked in relation to our exploration of educational purpose. This series of books can help establish a renewed interest in such questions and their centrality in the larger study of education and the preparation of teachers and other educational professionals.

More information about this series at http://www.springer.com/series/13431

Aristotelis S. Gkiolmas
Constantine D. Skordoulis
Editors

Towards Critical Environmental Education

Current and Future Perspectives

Springer

Editors
Aristotelis S. Gkiolmas
Department of Primary Education
University of Athens
Athens, Greece

Constantine D. Skordoulis
School of Education
National and Kapodistrian University
of Athens
Athens, Greece

Critical Studies of Education
ISBN 978-3-030-50608-7 ISBN 978-3-030-50609-4 (eBook)
https://doi.org/10.1007/978-3-030-50609-4

This Springer imprint is published by the registered company Springer Nature Switzerland AG
The registered company address is: Gewerbestrasse 11, 6330 Cham, Switzerland

I dedicate this book to the loving memory of my father Spyros, who taught me what justice is, to my beloved mother Katherine, and to Machi, Katia, Sotiris, and Spyros who fulfill my life with all the necessary…

Series Preface

Will We Be Still Here to *Share the Land?*

During a May Day trip into the Canadian Rockies, my daughter Meghann texted me that she was on the lookout for wild animals. Living near the mountains, we often see critters make their way from the hills, mountain goats, white tail deer, and on occasion a moose…indeed, deer are murderous predators in Alberta…the smashers and grabbers of gardens, the trouncers of landscape. Her search did not seem significant to me until she noted that the park rangers reported a sighting of a white bear. The woman who spotted the bear remarked, "*For us, it was a once-in-a-lifetime opportunity. Which is probably the only reason we decided to slow down and take a little bit of video of it…We took the opportunity to sort of educate him [her son] on, you know, bears and hibernation and berries and them keeping safe… it was great to talk about our environment and where we live and how we relate to them. And what better way to explain to him what it means to live with wildlife than to be able to see it,*" the mom was talking about environmental education and the unique opportunity to share the white bear (Gibson, 2020). While a sweet and encouraging story, the underlying theme is that we do not have acceptable environmental education and we do not have the words, the background, the examples, nor the scholastic resources to create an environmentally savvy populace.

The uniqueness of this nature scene unravels into the everyday and expands onto a global mural when we contextualize why the family was able to see wildlife in the wild. May 1, 2020, was somewhere close to day 61 of the coronavirus pandemic, and the *urban wild* was in its infancy (The Guardian, 2020). *The Guardian's* lead article covered mountain goats marching single file in Llandudno, Wales; joggers carefully avoiding possibly rabid coyotes during the day in Hayarkon Park in Tel Aviv, Israel; others passing jackals howling on the same trail that evening; a sea lion wailing in front of a closed store in Mar del Plata Harbor in Buenos Aires; and packs of sleek, gray langurs dancing in the middle of a highway in Ahmedabad, India… coincidence? Not at all…

In our little local world, we have been searching the neighborhood for our enormous jack rabbits. These are not Easter bunnies; these are dog-sized beasts…chameleons who by winter are pure white and by spring molt into brown. In *normal* life, I walk outdoors and am greeted by seven or eight of these feral lagomorphs, remembering the 1970s sci-fi film, *The Night of the Lepus*, wherein giant mutants terrorize the Southwest. But *normal* has changed…and like Israel, our new urban wild is rampaging coyotes as the surviving fittest who have gobbled up all of my rabbit friends. Clearly, the Earth's fauna was prepared to re-take the environment with aplomb.

The mom who used the white grizzly to teach her son while in the bear's own habitat was tapping into the notion of an ad hoc environmental education as on-the-ground, authentic pedagogy, respecting the ability for animals to be undisturbed, during the first stage of the coronavirus; it was the human inhabitants of Earth who were disturbed. And I cannot wonder, at the time of writing this *Foreword,* if the urban wild will survive or be pushed back into the threatened, disrespected wild of pre-virus times. Our world has been attacked, abused, and disrespected; and science and climate awareness had been disregarded in favor of greed, personal comfort, and ease.

The authors of *Towards Critical Environmental Education: Current and Future Perspectives* have created a volume which speaks to diverse cultures engaging in a global commitment to environmentally educate our children and youth. Reflecting on the 2020 COVID-19 pandemic, it is paralyzing to note that environmental education remains an add-on at best and a non-topic at worst in many schools. It is too early to tell if the pandemic will change curricula in any way, other than maybe the photogenic and fable-ized newsworthy shots of grizzly bears sunning themselves in the middle of parking lots in the urban wild. But the call from Gkiolmas and Skordoulis and authors shouts to us to awaken to the dire consequences of an environmentally ignorant world.

The contributors of this book have walked us through a critically informed pedagogy, which can be applied in both indigenous and Western constructs of education with the possibility of crossover and application, taking the best from geographical and ideological frameworks. The authors move environmental education into meta-world hopes for health, equity, justice, diversity, dialogue, and renewal. We have choices, at least right now…if, as educators, we understand the existential threat to humanity, flora, fauna, ideas, and thought is rooted in the environment…then we are ethically bound to insist on environmental education become foremost in our lives and vocation. One mom explaining a bear in the wild is not enough to infuse an indelible defense of environment for the ages. My opinion? If we do not fight to create a criticalized curricular space for the environment, well…no one will be around to ponder what *could have been done* by the time we would have reached the twenty-second century.

References

Gibson, J. (May 1, 2020). *Holy smokes…those are grizzly bears. And one of them is white*. Toronto, ON: CBC News. Retrieved from https://www.cbc.ca/news/canada/calgary/white-grizzly-banff-alberta-rare-1.5552431

The Guardian. (April 22, 2020). *The urban wild, animals take to the street amid lockdown—in pictures*. London. Retrieved from https://www.theguardian.com/world/gallery/2020/apr/22/animals-roaming-streets-coronavirus-lockdown-photos

Calgary, AB, Canada

Shirley R. Steinberg

Contents

About the Series Editor

Shirley R. Steinberg is Research Professor of Critical Youth Studies at the Werklund School of Education in Calgary, Canada. She is the author of many books, articles, and films on critical pedagogy, equity and social justice, anti-racism, and media literacy.

Chapter 1
An Ecocritical Conceptual Framework Toward *Eco*tistical Pedagogies

John Lupinacci

The past decades of environmental education (EE) scholarship have been marked by strong critiques of how neoliberal policies and reform efforts have contributed to an erosion of valuing the gravity of our human dependencies on the health of diverse species and ecosystems on the planet. In fact, we are living in frightening times in which ignoring the very real impacts of climate change have become an everyday part of global politics. However, environmental educators have long now maintained the importance of valuing and acting in defense of diversity as a core foundation of democratic life, and many EE researchers and environmental educators have committed to the possibilities of addressing the cultural roots of social justice and sustainability in a myriad of scholar-activist ways. Considering the stark conditions for life on the planet due to climate change, poverty, famine, and increased violent conflict, scholar-activist environmental educators are more than ever presented with the challenge of rethinking EE and doing so with close attention to what can be done differently. In this era, one in which global climate change is threatening the very existence human and more-than-human communities, it is from the position of an ecocritical teacher educator and EE researcher that I suggest we turn our attention toward the possibilities of radically reconsidering the purpose, and the associated procedures and practices, of not only EE but also of PreK-12 schooling in its entirety. In this chapter, I introduce an ecocritical framework for EE with a focus on working with teachers and recognizing that within the important relationships between EE, teacher preparation, and higher education, there exist the opportunity for critically addressing and rethinking current dominant conceptual frameworks constituting classrooms, schools, and communities. Furthermore, I share how anthropocentrism in connection with assumptions of human supremacy become a distinguishable focal point for ecocritical pedagogies. Concluding, I share some

J. Lupinacci (✉)
Washington State University, Pullman, WA, USA
e-mail: john.lupinacci@wsu.edu

A. S. Gkiolmas, C. D. Skordoulis (eds.), *Towards Critical Environmental Education*, Critical Studies of Education 14,
https://doi.org/10.1007/978-3-030-50609-4_1

actions that I have found to be helpful in taking steps toward what I propose—echoing diverse indigenous epistemologies—is a shift from egotism to *eco*tism enacting *eco*tistical pedagogies with(in) EE and teacher education.

Responding to the systemic violence and exploitation perpetuated by the current dominant social, economic, and environmental contexts in North America (similar to other nation states imbued in Western industrial culture), ecocritical environmental educators examine and address how it is that schools in Western industrial culture create, support, and sustain the habits of mind that rationalize, justify, and (re) produce unjust social suffering and devastating amounts of environmental degradation. When faced with such a challenge, ecocritical educators ask, *How is it that exploitation is rationalized, justified, and/or (re)produced through how we teach?* Coupled with that question, ecocritical educators and researchers in EE are also committed to turning the critical lens inward and asking: *What can EE researchers, educators, and activists do to teach in support of alternatives to Western industrial culture?* In an attempt to address these questions, I draw on collaborative work with diverse colleagues to introduce what is referred to as an ecocritical approach to exploring the possibilities of diverse critical ecological perspectives in EE research and teacher education.

An ecocritical approach addresses how education is influenced by systems of exploitation and violence, systems which rely on a refusal to acknowledge and embrace mutuality and interdependence (Lupinacci & Happel-Parkins, 2015). Ecocritical scholars use diverse critical lenses for addressing and rethinking dominant cultural frameworks, but certain principles remain at the center of the work. Specifically rooted in and yet pushing the boundaries of the critical tradition, teacher educators positioned within the ecocritical movement recognize that social and environmental justice are inseparable and inextricably linked, and that these injustices rely on the perpetuation of value-hierarchized social thought. In order to dismantle such injustices, then, we must analyze the culturally constituted value hierarchies our society has created. This approach also includes exploring alternative knowledges and ways of understanding difference that move beyond the limitations of Eurocentric (or Western industrial) thought.

Introducing Ecocritical Pedagogies

Ecocritical pedagogies in EE and teacher education is aimed at engaging teachers in identifying and critically examining the role that education both plays, and ought to play, in transitioning toward supporting diverse, socially just, and sustainable communities. Drawing from an EcoJustice Education framework (Martusewicz et al., 2015) and stemming from the growing field of ecocritical work in social and cultural foundations of education, I summarize the movement's aims in three aspects that frame ecocritical perspectives as working with teachers to critically and ethically:

1. Examine Western industrial culture and its impacts on social and environmental systems.
2. Examine value-hierarchized dualisms that contribute to inequities such as racism, classism, sexism, ableism, and anthropocentrism.
3. Examine and identify how to teach or share skills, and habits of mind, that support socially just and environmentally sustainable communities (Lupinacci & Happel-Parkins, 2016).

Simply put, through an ecocritical framework, teacher educators work to support scholar-activist educators in recognizing two conflicting and foundationally different worldviews—ecological worldviews of interdependence and interspecies equity and a dominant human-centered, capitalist, racist, ableist, heteronormative, and patriarchal worldview. Simultaneously, this framework shapes research in teacher education that also examines how those worldviews might be reconstituted—via education or ecocritical pedagogies—in ways that are local and in support of living systems (Lupinacci, 2013, 2015).

Ecocritical pedagogies often include students—together with their teachers: (a) recognizing that the role of teacher can be taken on by more-than-human members of any learning community; (b) examining how knowledge systems—in relationship to language, culture, and power—are culturally constructed; and (c) and how educators can play a role in reconstituting those relationships. By highlighting that our cultural belief systems, root assumptions, and narratives are constructed and not simply "natural," educators can help students develop critical perspectives on these root beliefs. This undertaking also opens the space for recognizing exploring alternative belief systems and metaphors that facilitate different kinds of relationships with other people, other beings, and the land.

An ecocritical framework also illuminates the systematic, economic, and political restructuring of lives that further perpetuates social suffering and environmental degradation. What has become commonplace over the past century, and extending into the current, is the intentional restructuring of relationships to control and commodify lives in order to maintain and manufacture markets. For example, food and water are life-sustaining elements necessary for supporting healthy communities. However, the relationships to these "resources" have been enclosed—monetized or understood as commodities to be earned and purchased. This iteration of capitalism—supply and demand economic systems predicated on exploitation—works to enclose living systems and can be understood as the globalizing force to commodify and privatize that which is common and public.

In short, ecocritical pedagogies center student learning on recognizing the importance of examining intellectual, environmental, and cultural practices and traditions in regard to how they either support or undermine living systems together with whatever content is being taught. Whether examining discursive practices or economic structures while learning mathematics, language arts, science, or social studies, a key feature of ecocritical pedagogies is the recognition that human knowledge systems are culturally constructed, have consequences for all living beings, and can be re-imagined in transformative ways (Turner, 2015). A distinguishing aspect of ecocritical pedagogies is that: Whatever the lesson or activity, students and teachers together are addressing the powerful role that their culture plays in the development of themselves, their values, and their diverse relationships. Such a framework examines, explores, and proposes diverse and collaborative pedagogical projects that respond to current dominant belief systems and works to ensure that any responses are necessarily collaborative with diverse cultures in ways that are local, situational, and in support of decentralized living systems.

A primary premise in ecocritical work that differentiates the approach from most other critical frameworks is that ecocritical educators assert that situated at the root of social and ecological injustice is a fundamental—and problematic—premise that humans, as a species, are understood (or self-identify) as superior to and somehow separate from all other living beings and nonliving things. Thus, guiding ecocritical pedagogies is the understanding that the manifestation of a human-supremacist worldview is culturally constructed and inextricable from current dominant cultural assumptions about race, class, gender, ability, age, and so forth. A foundational premise in ecocritical work in education is that cultural habits of mind in dominant Western industrial culture are based on a system of human supremacy—stemming from anthropocentrism—and that such a perspective is pervasive throughout how we as humans in Western industrial culture learn to interpret and assign value to differences.

The World Is Burning…So Why Should I Care

There are approximately 7.8 billon people in the world. Despite a growing awareness of human rights, the circumstances of an estimated 700 million people living in extreme poverty are dire (United Nations, 2014). Most often, women and children bear the brunt of this tragic, and at times fatal, suffering. The United Nations Children's Fund (2016) reports that "Unless the world tackles inequity today, in 2030: 69 million children under the age of 5 will die between 2016 and 2030" (p. 3), and the outlook for the world's impoverished youth is becoming increasingly grim. The United Nations (2015) reports that since 1990, global carbon dioxide (CO_2) emissions "have increased by over 50 per cent since 1990" (p. 8) and scientists have argued that this increase in CO_2 emissions is linked to changes in climate that contribute to floods, droughts, famine, and war (AMS, 2012; Andereeg, 2010; Doran & Zimmerman, 2009; IPCC, 2007, 2013; Oreskes, 2004). Given current conditions of social suffering, the increasing impact of climate change contributing to even greater widespread harm exacerbated by COVID-19, current education models need to be rethought, and their effectiveness evaluated in order to truly address the pressing issues of social justice and sustainability. Such matters have become life or death.

As the past decades of educational trends in EE and the impending crises we face in the next half-century clearly demonstrate, a refusal to understand and embrace mutuality and interdependence is woven throughout the interconnected hardships of social suffering and environmental degradation. This refusal is embedded in a conceptual framework based on a system of exploitation and violence—a lens that serves as the dominant, shaping force regarding what it means to be an educator. I assert that a dominant *egotistical* approach—or a form of teacher as a leader focused on the individual or self-development of students—must be examined and critiqued for its limitations in favor of efforts toward an *ecotistical* approach focused on the health and well-being of a broader ecological community and the interdependency and internationality of students as bodies of water, communities of bacteria, living breathing ecosystems within larger ecosystems. In confronting this stark contrast, I emphasize that as critical environmental educators, and especially as teachers in most school settings, we have a responsibility to examine and address how it is school cultures create, support, and sustain the extreme violence of social suffering and environmental degradation in the era now being called the Anthropocene. When EE scholar-activists are faced with such challenges, I reiterate that we must be willing to inquire into the ways that current forms of exploitation are rationalized, justified, and/or ignored.

Ecocritical Pedagogies in the Twenty-First Century

Central to an ecocritical framework is the importance of recognizing the differences between ecological cultures and intelligences and dominant human supremacist and individual-centered cultures. An ecocritical educational framework can be characterized as the examination and analysis of the ways in which culture, language, and the associated values and beliefs shape our thinking and contribute to injustice; as well as, our abilities and responsibilities to confront assumptions that underscore social suffering and environmental degradation. Such efforts require explicit attention to understanding diversity and the eco-social structural relationships between language, culture, and education that define how we recognize and understand difference. I draw from Weintraub et al.'s (2006) introduction of the term eco-tistical in response to a lack of terminology in the English language that describes "humans relating to the nonhuman environment in a harmonious, respectful, and pragmatic manner" (p. 55). By switching *ego-* to *eco-*, she intended to direct "focus away from self and toward home or habitat" in response to the absence of any opposite terminology for anthropocentric—the privileging of humanity—and egocentric—the privileging of self (Weintraub et al., 2006). In this sense, the term *ecocentric* is an adjective for ecological consciousness, or what Martusewicz and Edmundson (2005) propose as an "eco-ethical consciousness" (p. 73) that takes into consideration the social and environmental impact of decision-making and recognizes them as inextricably linked.

The primary emphasis of a growing body of scholarship identifying as ecocritical is a recognition of the limitations of how we understand and situate ourselves—humans—as a species that exists separate from and superior to all other forms of living and nonliving beings. Within that growing body of literature, scholars are increasingly using the phrase "more-than-human" (Abram, 1996, 1999) to draw attention to the larger set of ecological relationships that exist beyond those that imply merely humans in relationship with other humans. Thus, henceforth in this chapter, I use the term more-than-human to refer to plants, animals, streams, forests, soil, rocks, and so on to emphasize the existence of webs of connections that do not necessarily even include the human as a reference point. When addressing the injustices perpetuated by the conviction of human supremacy on and within the planet, ecocritical scholarship identifies the important role that schools, and more specifically educators and educational leaders, play in reproducing and reinforcing the root assumptions informing this belief. Therefore, an ecocritical framework is explicitly designed to interrupt and (re)constitute Western industrial assumptions that inform and structure how educators teach, and how this teaching shapes students' understanding of their human existence in relation to each other and, more specifically, to the living systems to which they belong.

As an ecocritical educator working to facilitate and encourage a change in mindsets for educators, I focus on *anthropocentrism*—or human-centered thinking. Anthropocentrism is the conviction that human beings are superior to other forms of life/matter and that human wants and needs supersede that of everything else. Importantly, I do not make the argument that anthropocentrism should take priority over other dominant discourses constructing Western industrial culture, such as patriarchy, racism, ableism, classism, and so forth. Rather, I assert that if a critique of anthropocentrism is too often not included in the critical examination of how we think as teachers, then it is likely we will continue to fail to address the deep habits of mind upon which many social justice issues are predicated. Our thoughts and actions all exist in complex relationships to one another, and our environments and the diverse relationships within them constitute and mediate how we construct meaning as a culture.

An important premise of an ecocritical framework is that the human-supremacist worldview is a cultural construct. In other words, this stance asserts that we, as humans—specifically those of us constituted by and constituting dominant Western industrial culture—have learned to think and behave according to culturally constructed, relational ways of understanding, and we use this understanding to interpret relationships and thus create meaning. Given the nature of meaning as culturally constructed, an ecocritical framework focuses on the ways in which meaning can be constructed in a manner that is supportive of the health and well-being of the entire community. Our cultural constructions can be interrupted and shifted when we learn to think differently about our relationships to each other and to the natural world. Thus, recognizing an anthropocentric worldview is an important entry point for rethinking human centrism and the role of educators to further teach toward (un) learning the injustice and pervasive violence of Western industrial culture.

An essential role of educators and researchers in EE committed to *ecotistical* pedagogies is to recognize and value teaching and learning that does not explicitly perpetuate human supremacy, and in so doing, work to identify and revalue the critical practices of valuing diversity, mutual aid, and interdependence that still exist in communities all over the world. To help explain this work, I draw from an image that compares a human-centered worldview to an ecological worldview.

Focusing on Fig. 1.1, I emphasize the need to engage in recognizing an anthropocentric worldview—that is, one that takes humans as the reference point of superior to everything Earthly—and how that worldview is culturally constituted and maintained. In the context of schools, educators and educational leaders can play a vital role in challenging this egocentric worldview that reflects our culture's anthropocentrism. EE researcher and educators have the ability—and arguably, the social and ethical responsibility—to partake in preparing a kind of global citizenry where diverse communities of humans both understand the need for and potential of (re) constituting this problematic and currently dominant worldview in favor of those more supportive of social justice and sustainability—such as an *eco*tistical worldview (pictured in Fig. 1.2).

Examining and discussing Figs. 1.1 and 1.2 provides an opportunity for dialogue about the ways in which worldviews can support or conflict with our educational efforts to address the challenges of the twentieth and twenty-first centuries. As EE educators and educational leaders engage in a critique of anthropocentrism, there is the opportunity to develop the capability to cultivate habits of mind which support and sustain all species, not just humans. These habits of mind help to challenge

Fig. 1.1 Egotistical Worldview adapted from "Differing Worldviews" (Martusewicz et al., 2015, p. 11). This figure illustrates the socially constructed hierarchical worldview of Western industrial culture. This illustration has appeared on a variety of social media networks and blogs—most notably the organization Generation Alpha. The Ego vs. Eco image—having made its way around through blogs, brochures, t-shirts, and posters—illustrates two fundamentally different worldviews. Adapting the Internet meme "Ego vs. Eco," Figure depicts an Egotistical worldview

Fig. 1.2 Ecotistical worldview adapted from "Differing Worldviews" (Martusewicz et al., 2015, p. 11). This figure illustrates an ecologically balanced worldview often overlooked and devalued in Western industrial culture.

destructive assumptions and practices promoted by anthropocentric egotistical leadership—a leadership approach that is all too prevalent and which often dominates school, family, and community politics. Recognizing the ways in which language influences culture and the ways in which culture influences language is essential as we conceptualize and implement changes to what could and ought to be truly inclusive communities. Thus, language plays an important role in how we interpret and examine the relationship between these two differing, and often conflicting, worldviews. For instance, EE conceptualized through an ecocritical framework challenges the prominence of dualistic and binary thinking in Western industrial culture. This process includes analyzing hierarchized superior/inferior dualisms in order to identify how such dualistic thinking works to uphold and perpetuate a problematic value-hierarchy that frames our understanding of our relationships with one another, ourselves, and the more-than-human communities to which we all belong.

Plumwood (2002) illustrates how in Western industrial cultures humans overwhelmingly understand relationality through sets of value-hierarchized dualisms. This thinking not only justifies and perpetuates anthropocentrism (human/nature), it also upholds forms of oppression such as racism (White/Person of Color), classism (Wealthy/poor), sexism (Male/female), ableism (Typical/atypical), and so on. All of these forms of oppression rely on value-hierarchized dualisms that inform how we understand and interact with one another.

To help explain how this applies to educational leaders learning to recognize sets of superior/inferior dualisms, I offer an example from my practice. The following is a list of some of the dualisms I use while working with teachers and educational leaders to address the logic structure of leadership in Western industrial culture. This list includes but is not limited to:

Superior/inferior
Central /marginalized
Human/nature,
Man/woman
Masculine/feminine
Reason/emotion
Mind/body
Wealthy/poor
White/Person of Color
Civilized/savage.
Master /slave.
Employer/ employee
Teacher/student
Adult/child

As we look at this list together, I ask educational leaders to stack some of the value-hierarchized dualisms together with superior/inferior and central/marginalized and then to work as a group to visually illustrate how these dualisms work within Western industrial culture to set up what is often referred to as the *norm* or the *standard.* Afterward, we highlight how these dualisms discipline our relationships, with an emphasis on the fact that while these dualisms inform how we think and act, it is important to remain mindful that "A map is *not* the territory it represents, but, if correct, it has a similar structure to the territory, which accounts for its usefulness" (Korzybski, 1933/1994, p. 58). Specifically focused on identifying and unsettling notions of superiority, we use this opportunity to examine how these hierarchized-dualisms combine to support exploitation and domination of others through our leadership and, conversely, how de(re)constructing them offers potentials for alternatives in how we lead.

To better understand the importance of de(re)constructing these dualisms, it is necessary to recognize how the terms and subject positions in the first column—which name those that are most highly valued within Western industrial culture—clearly illustrate how, for example, subject positions of wealthy individuals are considered more valuable than those in the second column, the poor. The high value and privilege afforded those in the first column are ascribed via a cultural logic structure that prioritizes and values them at the expense of those in the opposite column. This list illustrates how such dualisms combine for those identified as being the subject position within the left-hand column to reinforce and legitimize power over and control of those identified as subject positions in the column on the right.

Although such hierarchies are examined through critical pedagogies in teacher education and critical EE scholarship, the direct relationship of Human/nature to the other dualisms depicted in the list is a vital facet of this logic that is often overlooked. Pedagogically, I aim for these provocative placements, and specifically, listing them to illustrate how these value-hierarchized dualism function together, to encourage educators and researchers to critically and ethically question the fundamental assumptions in Western industrial culture about relationships when considering the work they set out to do and the support they intend to provide. I also intentionally use this list and the described process of drawing attention to the value-hierarchized dualisms and the connections between them to facilitate an understanding of the nature of these value hierarchies as intersectional. Further, I strive to cultivate an understanding of the ways in which these dualisms function—while acknowledging that they are not universally experienced—to expose the inequitable construction of what is often valued and rewarded as "normal" in Western industrial culture and which has manifested in an egotistical very male-centered, White, human supremacist leadership.

What Can We Do? Teaching Towards an *Eco*tistical Pedagogies in Education

With roots in critical pedagogy (Darder, 2016; Freire, 1993; Hooks, 1994; McLaren, 2006) and ecocritical pedagogical frameworks (Lupinacci, Happel-Parkins, & Turner, 2018; Bowers, 1993; Gruenewald, 2003; Furman & Gruenewald, 2004; Kahn, 2010), this work strives to identify and confront deep cultural assumptions informing worldviews in efforts to support educational leadership for social justice and sustainability. When we are faced with these commitments and the recognition that they are interrelated, it is important to recognize that our cultures play a significant role in how we think and act. It is from such a socio-linguistic and postmodernist position that educational leaders must learn to examine how and why we—as scholar-practitioners—think and act the ways we do. Accordingly, in educational leadership programs, students must learn to see the critical connections between empirical, social, and behavioral research and that they recognize their role as cultural workers in their communities (Freire, 1998).

As many issues of inequality and unjust suffering are embedded within educational structures that maintain and reproduce the unjust sociological phenomena of Western industrial culture, I strive to uphold the expectation that my teaching foster community-based learning that is rigorous, relevant, and builds strong community relationships. To support the development of *eco*tistical pedagogies in any community, teachers and educational leaders must become cultural workers, respected and disciplined researchers, effective and engaging speakers, and must both talk-the-talk and walk-the-walk.

Drawing from a diverse set of pedagogical frameworks rooted in critical theory, there is a strong connection between the conceptual framework and a commitment to the self-reflective process of sharing diverse understandings of what it means for a program to be committed to social justice and sustainability. With strong attention to the inextricable relationship between social justice and sustainability, this version of EE strives to recognize a that twenty-first-century challenges of social justice and sustainability require a strong commitment to understanding and interrupting the complex relationships that constitute, and are constituted by, dominant discourses and discursive practices of Western industrial culture in schools and society. Furthermore, it is essential that such interruptions be intricately and intimately intertwined with our own work in relationship to the tasks we ask of our students and future students.

Conclusion

In conclusion, the more that teachers, researchers, and educational leaders in EE engage in what this chapter proposes as an ongoing and necessary (re)constituting of EE and teacher education working together with indigenous educators and radical EE scholar-activists to continue to critically address anthropocentrism while moving toward the development and maintaining of an *eco*tistical pedagogical focus in EE; the more potential there will be for educational experiences in schools and communities for EE educators, researchers, and other leaders to foster spaces where teachers and students learn together to recognize the tendency of the privileged to dismiss what they would rather not confront. It is important to note that there are plenty of egotistical folks who for good reason ought to be admired and valued, and who are firm supporters of a shared commitment to respond to the undeniable atrocities that we—as humans—enact on one another. However, none of these atrocities occur in isolation and no solution or political (r)evolution will come from an egotistical authority. In confronting human supremacy and the egotism beholden to Western industrial culture's version of teaching and learning, it is paramount that critical EE leaders work as allies to all those suffering while challenging and confronting the systemic roots of oppression on all our respective fronts. In other words, we all have a responsibility—many of us as privileged members of society—to support those suffering unjustly in whatever capacity we can. As leaders striving toward an *eco*tistical society, we ought to be looking for any and all opportunities to decentralize power from individuals and facilitate a redistribution of power based on multispecies justice and equity.

(Re)constituting EE toward a focus on diverse *eco*tistical pedagogies requires EE educators, and researchers, in Western industrial culture to stop asserting leadership or authority in an egotistical manner, to turn attention toward the difficult necessity for shifting worldviews, and to committing to a cultural change that will mean giving up the power and privilege afforded to some at the cost of others and the environment. As EE researchers who identify as nonauthoritarian scholar-activist educators deeply embedded in Western industrial culture, for many of us this means learning to listen and practicing humility while trusting that if we, enactors of dominant Western industrial culture, do not (re)constitute the cultural framework by which dominant meanings are socially constructed, then we are destined to recreate many of the problematic relationships that we, as radical educators, often set out to change. I end this chapter with a list of suggestions aimed toward supporting and addressing the pervasive egotistical logics in education and turning our efforts toward more *eco*tistical projects.

- Engage in learning from the diverse projects in our human and more-than-human communities and commit to rethinking the dominant assumptions influencing how we, as humans, construct meaning and thus how we learn to relate to each other and the more-than-human world. Further, make the commitment to critically and ethically examine how we understand educating, organizing, and taking action toward supporting healthy communities that include all beings and the intrinsic value of recognizing, respecting, and representing the right of all beings to belong to and live in peace within healthy ecological system.
- Engage in critical and ethical examinations of what it means to be an educator, researcher, leader, etc. As notions of such are all too often defined in terms of human-centered egotism, it is important to work to (re)constitute these roles and responsibilities in terms of what and how an *eco*tistical pedagogy might facilitate as practice, and how those often competing but coexisting worldviews contribute to our actions either supporting or undermining the rights of all beings, including future generations, to coexist in peace.
- Engage in examining EE in terms of ecological systems and the diverse ways in which our living relationships can be recognized, respected, and represented through teaching and learning among all members. Specifically, engage in recognizing the role of activist networks in modeling diverse examples of how an *eco*tistical approach to leadership plays a role in alleviating and eliminating unjust suffering in our communities. In doing so work to recognize such resistance in diverse species. Powerfully strong examples in the United States can be found in human communities in the #Blacklivesmatter and Idle No More movements, as well as in more mainstream political movements like the Alexandria Ocasio-Cortez's campaign platform for a political revolution of the US government, or the Spore Liberation Front, which recognizes the importance and power of mycelial networks. In all cases, build networks of solidarity with these organizations. Furthermore, ask leaders in these movements about their visions and plans for education that supports social justice and sustainability.
- Engage in supporting the diverse approaches to leading and healing from Western industrial culture and in solidarity show respect for epistemologies that differ from the current dominant way of way of understanding and organizing learning. Support the ways in which diverse forms of resistance work to challenge and break the will of their oppressors and as leaders support this resistance even when it means giving up privilege and power.

- Engage in strong alliance with all those suffering and support the oppressed in solidarity while simultaneously working to shift and challenge the dominant systems that often perpetuate the suffering of marginalized and subjugated beings. In all cases as leaders show up, speak out, and take action to stop the systemic domination of one another, ourselves, and our more-than-human kin.

In such volatile and authoritarian times, it is important that ecocritical educators learn to work together to challenge dominant perceptions of what currently constitutes EE and work to (re)constitute an EE teaching and learning that takes serious social justice and sustainability. Efforts toward a political (r)evolution in education can begin by focusing on the importance of convivial EE networks—rather than on individual advancement. Through fostering the development of networks of scholar-activist educators studying and researching twenty-first-century challenges that include ecocritical dialogue, we can resist the overwhelming egotism and anthropocentrism in Western industrial culture and reject the illusion that as humans we are separate from and superior to each other and all other beings on the planet. We challenge egotism and anthropocentrism when we build solidarity in addressing common dilemmas in our communities with other *eco*tistical humans and especially when we teach one another to make choices to include in our educational networks more-than-humans—be they animals, mountains, trees, a river, the salmon that swim upstream to spawn, the large winged osprey that visits the lakes where we swim and cool off on a hot summer's day, the food that we grow, or the vast networks of mycelium in the soil. It matters mostly, in this sense, that we work away from understanding ourselves as independent individuals toward the kind of *eco*tistical understanding of self as interrelated and interdependent on the diverse living systems to which we belong. The point is that we learn a deep respect for difference and our shared dependencies when we understand in an ecological sense what it means to be ecotistical leaders—to recognize and value that we are in relationship with a diversity of wonderful beings and that we owe our existence to these devoted networks. From such relationships, we learn what it means to teach and lead by belonging and without framing that understanding within anthropocentrism; rather, belonging and respect become guiding EE principles that we enact in our everyday lives as *eco*tistical educators, researchers, and leaders. It is through these convivial and mutually sustaining relationships that we learn to overcome the isolating ills of Western industrial culture and our habits of egotism, and we are called to action with our diverse sisters and brothers to teach and lead in support of living systems. We practice *eco*tistical pedagogies.

References

Abram, D. (1996). *The spell of the sensuous: Perception and language in a more-than-human world*. New York, NY: Vintage Books.

Abram, D. (1999). A more than human world. In A. Weston (Ed.), *An invitation to environmental philosophy* (pp. 17–42). New York, NY: Oxford University Press.

American Meteorological Society. (2012). *Climate change: An information statement of the American Meteorological Society*. Washington, DC: American Meteorological Society. Retrieved from https://www.ametsoc.org/ams/index.cfm/about-ams/ams-statements/statements-of-the-ams-in-force/climate-change/

Andereeg, W. R. L. (2010). Expert credibility in climate change. *Proceedings of the National Academy of Sciences, 107*, 12107–12109.

Bowers, C. A. (1993). *Education, cultural myths, and the ecological crises: Toward deep changes*. Albany, NY: State University of New York Press.

Darder, A. (2016). Critical leadership for social justice: Unveiling the dirty little secret of power and privilege. *The Radical Imagine-Nation: The Journal of Public Pedagogy, 1*(1), 41–73.

Doran, P. T., & Zimmerman, M. K. (2009). Examining the scientific consensus of climate change. *Eos Transactions American Geophysical Union, 90*(3), 22.

Freire, P. (1993). *Pedagogy of the oppressed* (revised 20th anniversary ed.; M. B. Ramos, trans.). New York, NY: The Continuum Publishing Company.

Freire, P. (1998). *Teachers as cultural workers: Letters to those who dare teach* (D. Macedo, D. Koike, & A. Oliveira, Trans.). Boulder, CO: Westview Press.

Furman, G., & Gruenewald, D. (2004). Expanding the landscape of social justice: A critical ecological analysis. *Educational Administration Quarterly, 40*(1), 37–76.

Gruenewald, D. A. (2003). The best of both worlds: A critical pedagogy of place. *Educational Researcher, 32*(4), 3–12.

Hooks, B. (1994). *Teaching to transgress: Education as the practice of freedom*. New York, NY: Routledge.

IPCC (2007). IPCC forth assessment report: Climate change. Retrieved from https://www.ipcc.ch/assessment-report/ar4/

IPCC (2013). IPCC Fifth assessment report: Climate change. Retrieved from https://www.ipcc.ch/assessment-report/ar5/

Kahn, R. (2010). *Critical pedagogy, ecoliteracy, and planetary crisis: The ecopedagogy movement*. New York, NY: Peter Lang.

Korzybski, A. (1933/1994). *Science and sanity: An introduction to non-Aristotelian systems and general semantics* (5th ed.). Brooklyn, NY: Institute of General Semantics.

Lupinacci, J. (2013). Eco-ethical environmental education: Critically and ethically examining our perceptions of being human. In K. Young, A. Kulnieks. & D. Longboat (Eds.), Contemporary Studies in Environmental and Indigenous Pedagogies: A Curricula of Stories and Place (pp. 185–200). Rotterdam, Netherlands: Sense Publishers.

Lupinacci, J. (2015). Recognizing Human-Supremacy: Interrupt, Inspire, and Expose. In A. Nocella II, K. Socha, R. White, & E. Cudworth (Eds.) Anarchism and Animal Liberation: Essays on Complementary Elements of Total Liberation. (pp. 179–193). Jefferson, NC: McFarland.

Lupinacci, J., & Happel-Parkins, A. (2015). Recognize, Resist, and Reconstitute: An Ecocritical framework in Teacher Education. *The SoJo Journal: Educational Foundations and Social Justice Education, 1*(1). 45–61.

Lupinacci, J., & Happel-Parkins, A. (2016). Ecocritical foundations: Toward social justice and sustainability. In J. Diem (Ed.) The Social and Cultural Foundations of Education: A Reader (pp. 34–56). San Diego, CA: Cognella

Lupinacci, J., Happel-Parkins, A., & Turner, R. (eds.) (2018). Ecocritical scholarship toward social justice and sustainability in teacher education. *Issues in Teacher Education, 27*(2)

Martusewicz, R., & Edmundson, J. (2005). Social foundations as pedagogies of responsibility and eco-ethical commitment. In D. W. Butin (Ed.), *Teaching social foundations of education: Contexts, theories, and issues* (pp. 71–92). Mahwah, NJ: Lawrence Erlbaum Associates, Publishers.

Martusewicz, R., Edmundson, J., & Lupinacci, J. (2015). EcoJustice education: Toward diverse, democratic, and sustainable communities (2nd Edition). New York, NY: Routledge.

McLaren, P. (2006). *Life in schools: An introduction to critical pedagogies in the foundations of education* (5th ed.). Upper Saddle River, NJ: Pearson.

Oreskes, N. (2004). Beyond the ivory tower: The scientific consensus on climate change. *Science, 306*(5702), 1686.
Plumwood, V. (2002). *Environmental culture: The ecological crisis of reason*. New York, NY: Routledge.
Turner, R. (2015). Educating for ecojustice: Curriculum and lessons for secondary and college classrooms. New York, NY: Routlege.
United Nations. (2014). *The millennium development goals report 2014*. New York, NY: United Nations. Retrieved from http://www.un.org/millenniumgoals/2014%20MDG%20report/MDG%202014%20English%20web.pdf
United Nations. (2015). *The millennium development goals report 2015*. New York, NY: United Nations. Retrieved from http://www.un.org/millenniumgoals/2015_MDG_Report/pdf/MDG%202015%20rev%20(July%201).pdf
United Nations Children's Fund. (2016). *The state of the world's children 2016: A fair chance for every child*. New York, NY: United Nations Children's Fund. Retrieved from https://www.unicef.org/publications/files/UNICEF_SOWC_2016.pdf
Weintraub, L., Phillips, P. C., Smith, S., Godfrey, D., Vesna, V., Weintraub, L., et al. (2006). Forum: Eco-tistical art. *Art Journal, 65*(1), 54–81.

Chapter 2
Paulo Freire and a Curriculum for the Capitalocene

Samuel Day Fassbinder

> *"The future does not make us. We make ourselves in the struggle to make it."—Paulo Freire (Pedagogy of Indignation, pp. 34).*

Introduction: A Curriculum for the Capitalocene

The idea of the Capitalocene comes out of a proliferation of recently created concepts (see especially Fressoz and Bonneuil's *Shock of the Anthropocene* for a glossary of them, in which one can read of the "thanatocene," the "phagocene," and so on), all of which are used to dramatize the human species' impact upon the environment. The original concept of all these is the Anthropocene—a concept used to integrate human history into natural history by specifying an epoch, a time division, of natural history marked by human alterations of the landscape. The "Anthropocene," then, is the geologists' and paleontologists' claim to be able to say what human history is about. For those scientists, this epoch, the epoch of human history, is about the vast infrastructure covering the planet.

Therefore, today, right now, we are "in" the Anthropocene. When did that epoch begin? Dates vary, but an oft-used year for the "Anthropocene" is 1945, the last year of World War II and the year in which nuclear explosions began spreading radioactive dust around the world. A significant statistical argument has been put together to show how human impact upon the environment became a massive, and accelerating, endeavor, using 1945 as a starting point for what are presumed to be the really big moves. This argument is summarized in McNeill and Engelke's (2014) book *The Great Acceleration: An Environmental History of the Anthropocene since 1945*. Perhaps the most obvious indicators of the "Great Acceleration," thus the title, are the progressive increase in total human population—from two billion in 1930 to

S. D. Fassbinder (✉)
DeVry University, Naperville, IL, USA

A. S. Gkiolmas, C. D. Skordoulis (eds.), *Towards Critical Environmental Education*, Critical Studies of Education 14,
https://doi.org/10.1007/978-3-030-50609-4_2

more than seven billion in 2011 or 2012 (41)—and the progressive extraction and combustion of fossil fuels, with subsequent increases in the Earth's atmospheric carbon dioxide content. (This summary will skip the discussion of the proliferation of infrastructure, with its lighting-up of the night sky through the vast proliferation of streetlights in every city in the world, the coating of the Earth's surface with roads and cities, and so on.)

The "Anthropocene"—as a topic—is now part of a number of different discussions of the environment and human history. The environmentalists, historians, sociologists, and other social thinkers no doubt wish to claim a piece of what the geologists appear to be claiming for themselves, since this is a geologic period encompassing human history. But what characterizes the "Anthropocene" discussion, more than anything else, is the notion that natural history is no longer separate from human history.

One way of figuring out what claim the social thinkers have upon the "Anthropocene" concept is through an exploration of the "Capitalocene" concept. Perhaps the term "Capitalocene" is just mimicry, inserting part of the word "capitalism" into what was otherwise going to be a period of natural history. The idea of the "Capitalocene" is to say: "Ha! It's capitalism, and not just people per se, that changed the Earth." Or, more compactly, the Capitalocene "signifies capitalism as a way of organizing nature" (Moore 6) as Jason W. Moore describes it.

But the vast changes in life on Earth that have occurred since that long-ago time when agriculture was invented, since 1492, the year of the Spanish conquest of the New World, since 1882, the year in which the large-scale power plant was invented, or since 1945 cannot be imagined as mere geological facts. Such changes are not mere byproducts of human domination of the Earth. The simple story, which is that human beings became the dominant large mammal over every part of the land surface of the Earth, omits discussion of the much larger time period in which the human impact upon the rest of planet Earth was much less forceful than what it is now. The human species is at least 200,000 years old; the age of massive impact upon planet Earth is a rather recent phenomenon when regarded within that timescale. So for instance the extraction of tens of millions of barrels of oil (a barrel is 42 gallons) *every day* (and its subsequent burning) was something that didn't happen until recently. Statista.com tells us: "Worldwide, around 94.7 million barrels of oil were produced daily in 2018."[1] Such massive environmental phenomenon drama is a recent phenomenon, unique to history.

If we are to look at the Great Acceleration, and thus also at the Capitalocene, as a contribution to some imagined future's fossil record, the first thing that the future fossil-diggers will note is the present-day extinction event. For paleontologists, the fossil record, locked in geological strata, is punctuated by occasional extinction events, points in the record traceable to prehistoric times in which certain fossils no longer appear. So, for instance, at a point in this record dated to 66 million years ago, dinosaurs, ichthyosaurs, plesiosaurs, ammonites, pterosaurs, and toothed birds

[1] https://www.statista.com/statistics/265203/global-oil-production-since-in-barrels-per-day/

disappear from the fossil record; this is known as the "end-Cretaceous extinction event." We are living through such an extinction event right now; a UN report of 10 years ago speculates that 150–200 species of plant or animal go extinct every day (Vidal).[2] As Ashley Dawson (2016) points out, this event can be traced to capitalism and is contemporary with Europe's conquest of the world since 1492. To be sure, there were two previous extinction events of importance within the human species' existence—the "end-Pleistocene extinction event," in which megafauna such as the woolly mammoth disappeared, and the local extinction event that accompanied the rise of the Roman Empire, in which enormous numbers of large animals were put to death in cultural rites (Dawson 20).

As a response to the extinction crisis, Dawson suggests an appeal to creative anti-capitalism:

> It has been said that it is easier to imagine the end of the world than to envisage the overthrow of capitalism. I would respond to this aphorism from dark times that it is easier to imagine the end of capitalism than it is to articulate any other genuine solution to the extinction crisis. If capitalism is the ultimate cause and prime engine of the extinction crisis, surely we can only conclude that we may find hope in challenging its baleful power with all means at our disposal. (57).

In sum, there is a social cause, a social force enacting a cultural mega-event, impacting the human race and its environmental substrate, responsible for the "Great Acceleration." Its name is capitalism. It counts as an extinction event, but the extinction perspective is just one perspective upon the whole drama of present-day life, there also being a general environmental crisis encompassing deforestation, pollution, climate change, the harmful side-effects of conventional agriculture, plastics in the oceans, and so on. Of these damages, climate change is probably the most insidious, as it will very likely intensify over the next few decades, with possible famine, plague, drought, rising sea levels, and consequent refugee crises on the horizon. The term "Capitalocene," among all the others, names the primary social force changing everything, positioning it as a big deal within the whole of natural history. The "Capitalocene" prompts this one question: so capitalism has changed the world, impacted planet Earth as no other social force has done throughout all of human history/ prehistory. What do we do now?

This question is also specifically a question for educators. What do we do now? As a response, this chapter proposes a "curriculum for the Capitalocene," a college-level educational response to the Capitalocene, with its "Great Acceleration." A first consideration of such a curriculum is to spread awareness of what is happening now to display the current period as a rather exceptional time not only in natural history but also in human history. That, within any educational framework, is the subject matter. To ask "what do we do now," however, is to motivate students to ask and to pose the current period as a problem—for this we will need a "problem-posing" education, and so the historical educator Paulo Freire (1921–1997) will be our guide.

---

[2] https://www.theguardian.com/environment/2010/aug/16/nature-economic-security

One way of approaching the "what do we do now" question is through the concept of ecopedagogy of education for planetary citizenship. We can see this in Greg Misiaszek's (2018) book *Educating the Global Environmental Citizen*. Misiaszek, however, takes his cue from the "citizen schools" that were the focus of Paulo Freire's thinking toward the end of his life (Gadotti & Torres, 2009). Misiaszek sees his task as one of "teaching global citizenship" (18). This is ostensibly to be done through the teaching of "critical environmental literacies" (Chap. 1), "environmental citizenship" (Chap. 2), "theoretical lenses" (Chap. 3), and "development" (Chap. 4), in what can be read as a set of guidelines for a possible curriculum. This is Misiaszek's definition of "planetary citizenship":

> Planetary citizenship is a reconstruction of how we view ourselves with all else that makes up the planet, including the atmosphere, as a single being, as Earth as a citizen. Explained in detail throughout this book, this citizenship sphere allows for biocentric lenses of viewing environmental ills, that is beyond anthropocentric perspectives that center humans' interests, whether they coincide with socio-environmental justice or not. In short, the planetary citizenship sphere provides for the widest perspective of what is environmental justice with solidarity of all that is Earth, including ourselves as part of Earth, but without singularly centralizing humans' socioenvironmental justice. (18).

Misiaszek's way is one way of doing it. This chapter suggests another option, one not predicated upon prior conveyance of a "biocentric" lens, and one conditioned by a North American experience of college life and focused upon two concepts: "praxis," and "utopia." More specifically, the perspective here suggests an openness to praxis and to utopian dreaming as conditioned by an understanding of the current moment as dominated by the acceleration of an out-of-control Capitalocene, the great expansion of humanity and production and technology and infrastructure of recent times. Freire argues that "praxis," or "reflection and action directed at the structures to be transformed" (1993/1970. 107), is an important goal of education; but more importantly, Freire was beginning to contemplate an ecopedagogy toward the end of his life. Freire's life took place at a time in history in which there was an ongoing Age of Utopian Dreaming, in which Freire enthusiastically participated. The utopian dreams of that age, more or less between the Enlightenment and the 1970s, inspired people to look for positive alternatives to existing social structures rather than thinking within the confines of neoliberalism. It is this past experience which will be recruited to inspire the activism of the present day.

For Freire, the student was a participant in adult literacy; in another sense, the student was an activist-in-the-making. One can discern from "reading the word and the world," Freire's first principle of literacy, that two concepts were being bundled: learning to read, and advancing one's interpretations of how things are. Freire taught adult literacy; but beyond the mere characterization of the student as a reader and as a writer, she was a potential transformer of the world.

Freire thus offers a heroic-potential conception of the student, which in other passages is used to endorse a *particular* transformation of the world. Freire's conception of the student nonetheless begs the question for today's readers of how the world is to be transformed. More topically for today, the praxis that goes on in and beyond the Freirean classroom should be about creating an ecotopia, a utopia of sustainability. A utopia of sustainability would resolve the environmental crisis by concluding the Capitalocene, rather than said crisis being an ongoing (and eventually losing) battle against capitalist agents of environmental harm—mining interests, sellers of poisonous chemicals, capitalists in the warfare business, and so on.

Utopian dreaming becomes necessary to this particular curriculum for the Capitalocene because any realization of the relevance of the "Capitalocene" concept will have to come out of the student perspective. A focusing statement of Freirean pedagogy can be found in the Second Letter of *Pedagogy of Indignation*, in which he discusses how it is not possible to transform the world without a concept of "utopia" (31). Freire says, "The future does not make us. We make ourselves in the struggle to make it" (34). The student, then, is potentially someone who makes the commonly experienced future, and "utopia" is the motivating concept to inspire the imagination of a better future than the experienced present moment.

In the secondary literature on Freire, the high point to be reviewed here will be the Misiaszek book and Georgios Grollios' (2016) *Paulo Freire and the Curriculum*, but, beyond Misiaszek and Grollios, the curriculum at the bottom of this essay will engage the Capitalocene concept. Grollios' book is invaluable for the connections it makes between what Freire actually did and the historical circumstances of the times in which he lived. Late in life, Freire looked at the notion of ecopedagogy. In the framework below, ecopedagogy would start with the fantasy of a sustainable society, a society that can survive the environmental dislocations of the Capitalocene. It is in this spirit that a curriculum appropriate to the Capitalocene will be proposed.

The proposed curriculum will offer students an ecopedagogy. Misiaszek (2018) offers one definition of ecopedagogy:

> Reinvented from the work of Paulo Freire (2004, 2000), ecopedagogy is grounded on action-oriented teaching through democratic dialogue to better understand how environmental ills oppress people, societies, populations, and everything on the planet. (14).

This definition fits the model granted in Freire's *Pedagogy of the Oppressed* of teaching as activism through democratic dialogue within a dichotomy of liberation and oppression. It's an essential attempt to cope with the circumstances of the Capitalocene.

Generally, though, a particular curriculum will specifically depend upon what teacher and students know how to do, what the classroom situation needs, what course is ostensibly being taught, and what are the historical forces at work in the society inhabited by teachers and students; here there will be some general guidelines. Paulo Freire had answers for his times and places and was thus able to say what curriculum was for their teaching situations; we will need answers for ours. In each instance, though, the appeal of a Freirean curriculum was in an opening up to a utopian dream, the dream of a better world in which other, less obviously utopian, dreams would be rendered possible.

Freire's short book *Cultural Action for Freedom* suggests a direct approach to utopia in Freire's idea of pedagogy. An essay by Darren Webb, "Educational Archaeology and the Practice of Utopian Pedagogy," summarizes:

> The first use of the phrase "utopian pedagogy" comes in 1970, in Paulo Freire's *Cultural Action for Freedom*. Here he describes utopian pedagogy as a process of denunciation and annunciation; a critical interrogation of the present situation coupled with "a utopian vision of man and the world" (Freire 1972 Freire, P. 1972. *Cultural Action for Freedom*. Harmondsworth: Penguin. [Google Scholar], 40).

The Freirean educator works, then, to put the utopian dream into play in the lives of her students.

In the concrete world depicted by history, Paulo Freire (1921–1997) was an educator who had significant success in teaching lower class Brazilians how to read and write, and whose thought was picked up by a number of literacy programs in developing countries. His influence persists to this day through his foundation in Brazil and through "critical pedagogy," a subfield of educational practice in which he can be imagined as the most important historical figure. To some extent, Freire came to the history of education in the right place and at the right time; developing countries in the era of the Cold War were interested in securing his endorsement of their literacy programs for political reasons, and in such a way, he achieved international fame.

Freire's literacy-teaching techniques (see Taylor, 1993 for further discussion), his philosophies of critical pedagogy, practices of problem-posing education, and curricula of generative themes may have inspired a number of students in Freirean-based programs to change the societies in which they lived.

Before the problem of a "Freirean curriculum for the Capitalocene" can be solved, however, the problem of educational context must be solved. The trend with curriculum, as the Internet expands, is to move classes online: but as William F. Pinar suggests, "when curriculum is moved online, students can remain at home, and teachers, with whom students may have no direct contact, act as auditors, with commensurately reduced wages and intellectual influence" (2013:16). Online classes within the traditional framework of grades, credits, and diplomas suggest a shrinking of the fantasy content of the teacher; the "industrial-consumer" model (which privileged the few while subordinating the many; see Joel's Spring's *Pedagogies of Globalization* for an explanation of this model) is to be replaced by an Internet-based "self-help" model, which reproduces the systems of education which were there before the Internet, inequitable as they were, but which today appear more grown-up.

Most useful as a guide to educational content is the portion of the Grollios book (mentioned above), which discusses Freire in the 1980s and 1980s. During this time Freire was, briefly, a Secretary of Education for the City of Sao Paulo. Even though Freire's focus had shifted from critical pedagogy in the context of literacy programs in developing nations to a critique-from-within of education-as-it-is, which remains appropriate for the project proposed here. In this later stage of Freire's life, what is radicalized first is the school situation: "To Freire, school continued to be a place of social conflict. In other words, class struggle still took place inside the school" (Grollios 89). Praxis, therefore, was constituted by radicalized notions of what disciplinary teaching was:

> As much as Freire considered it certain that a biology teacher should teach biology, he considered it equally certain that the phenomenon of life could not be deliberated without

> discussions of exploitation, domination, freedom, and democracy. The content of a scientific subject could not be separated from its historical and social context. As the teachers should not leave content in parentheses and simply discuss the political situation of the country with their students, they should likewise not put history and social conditions in parentheses, thereby making it clear to their students that there is no such thing as biology per se. (90).

Thus for Freire the reality of the educational situation must be brought into the curriculum, conceived in a larger context, obliging the instructor to teach a subdiscipline: political physics, political biology, political chemistry, and so on. (I am using the word "political" here for lack of an alternative adjective which isn't depoliticized. Otherwise "participatory" would be a more appropriate term.) For our purposes, the subdiscipline will combine a knowledge of history, specifically a history of Utopian Dreaming, with an empowering participation in utopian dreaming of the students' choosing.

The next portion of this essay will then look critically at options for praxis. It is one thing to place utopian dreams onto the school menu, and another to place changing the world onto that same menu. First, the "deschooling philosophy" of Ivan Illich (1926–2002), offering a collective and utopian "self-help" philosophy appropriate to an age of the Internet. Deschooling philosophy will be examined as providing options for praxis and for utopian dreams. Illich appears hardheaded and practical in a reading from today's era of the Internet, but the real-life Illich was insufficiently willing to change the world so as to realize his utopian vision. Second, there will be a critique of the idea of "antistructure" in the classroom as portrayed in Peter McLaren's (1993/1986) *Schooling as a Ritual Performance*. As institutions, schools tend to promote a give-and-take between everyday rebellion and institutional conformity, which favors conformity. Appearances of antistructure are brought in line with an ongoing status quo of social reproduction. It becomes up to the reader of McLaren's text to identify how antistructure can benefit needed goals of social change.

Two Models of the Student: Illich and McLaren and the Possibility of Praxis

Ivan Illich merits mention here because he was, like Freire, a product of the educational utopianism of his time, most notably the 1960s. Illich's deschooling philosophy, promoted most famously in his (Illich, 1970a) *Deschooling Society*, suggested the student's independence from schools and teachers. In opposition to school as an institutional site of learning, Illich favored the "learning web," which he defined as the "inverse of school," and which he based on the principles "that we can depend on self-motivated learning instead of employing teachers to bribe or compel the student to find the time and the will to learn" and "that we can provide the learner with new links to the world instead of continuing to funnel all educational programs through the teacher." With the promotion of the learning web, education disappears as an academic field, replaced in its stead by subject matter. Illich's (1973) book *Tools for Conviviality* displays his utopian dream in greater detail. In that volume, Illich dreamed of a "society in which modern technologies serve politically interrelated individuals rather than managers" (xxiv). This is the dream of a society without institutional constraints, in which society would be "reconstructed to enlarge the contribution of autonomous individuals and primary groups to the total effectiveness of a new system of production designed to satisfy the human needs which it also determines" (11). This is a basic ecotopian ideal—production for human need rather than for the inflated "needs" of the consumer society.

The ideal suggested in *Tools for Conviviality* was that technologies, from the most basic to the most sophisticated, should be designed to be "hackable" by individuals rather than as instruments of social control. School, in this formulation, is the first institution we encounter, which places our fate in managerial hands, and so Illich says, "the transformation of learning into education paralyzes man's poetic ability, his power to endow the world with his personal meaning" (60). A deschooled society, then, would be society in which students were not present for the "progressive consumption of curricula" (62) but rather for their own collective "poetic ability."

There is indeed a straightforward presentation of Illich's utopianism, in an essay titled "A Call to Celebration" (Illich, 1970b). It calls for the "end of privilege and license": "we are challenged to break the obsolete social and economic systems which divide our world between the overprivileged and the underprivileged" (15–16). It is a call to "live the future": "let us join together to celebrate our awareness that we can make our life today the shape of tomorrow's future" (18). This is the utopianism of empowered people, people who can make a difference, and who are thus prepared to bring the Capitalocene to an end and bring into being a new, sustainable society – if it were possible. Illich's utopianism is almost implicit—it assumes the wish "to use mankind's (sic) power to create the humanity, the dignity, and the joyfulness of each one of us" (15). The question with Illich, as with others of pure-hearted expression, is one of how do we get from what we have to the utopia we could have.

There must, however, be a prior democratic social existence not fleshed out by Illich if Illich's utopian dream were to come true, something facilitating the "union of free producers" that Illich wanted to spread to education. This appears in the idea of "skill exchanges" in *Deschooling Society*:

> A much more radical approach would be to create a "bank" for skill exchange. Each citizen would be given a basic credit with which to acquire fundamental skills. Beyond that minimum, further credits would go to those who earned them by teaching, whether they served as models in organized skill centers or did so privately at home or on the playground. Only those who had taught others for an equivalent amount of time would have a claim on the time of more advanced teachers. (90).

This brings to mind Karl Marx's description of the first stage of communism in the "Critique of the Gotha Program." Marx:

> What we have to deal with here is a communist society, not as it has *developed* on its own foundations, but, on the contrary, just as it *emerges* from capitalist society; which is thus in every respect, economically, morally, and intellectually, still stamped with the birthmarks of the old society from whose womb it emerges. Accordingly, the individual producer receives back from society – after the deductions have been made – exactly what he gives to it. What he has given to it is his individual quantum of labor.

The skill exchange appears in this sense as an "initial stage of communism" for learners, conceived within a social order which rewards their original contributions.

What's important to Illich's vision of the student is that the student be independently ready to learn. There are, in real life, schools which socialize independent learners—A.S. Neill's model of schooling in the book *Summerhill* describes a school which was (and is; Summerhill still exists, near the east end of England) not oriented toward accumulating knowledge or abilities but rather a common "learning through freedom" philosophy in which students would eventually be obliged to choose a learning path that would allow them to have a life after school. Mostly, however, the student choice of a learning path in such schools is coerced by the necessity of the student "having a life" (in institutional terms) after the years of mandatory schooling have passed.

What will coerce students working with a "curriculum for the Capitalocene" will be the prospect of life within the Capitalocene—the travails of the capitalist system, the constant reshaping of the Earth by the forces of production, the consequent environmental crisis. The resultant questions will focus upon utopia, which is to say, about how desire and imagination are projected onto the future, and praxis, which suggests the question asked above of what do we do. The "curriculum for the Capitalocene," then, is assigned the task of bringing all of the above to consciousness.

The immediate question suggested by readings of Ivan Illich, in this regard, should be about the survival of independent learners in the context of present-day life. Independent learners are somewhat hardy—they can survive in numerous places and be part of numerous collectives, not all of them empowered or empowering. Independent learners might be "passed along" in traditional schooling structures by professors who respect their initiatives; but said learners might otherwise have very little use for a number of other architectures within the school. Exam scores and diplomas publicize narrowly focused types of knowledge, but don't apply it; more appropriate for the independent learner would be exhibitions or portfolios or projects. Letter grades impose exchange-value upon the learning process and are anathema to the Illichian student, who learns because she wants to know something (and not because some external reinforcement is being offered her).

Perhaps the American university could be made into an easier place for the Illichian student. It would have to happen after the commodity pressure is removed from university life through, for instance, the "College for All" Bernie Sanders proposal. The Illichian student operating without "College for All," however, might probably find a more agreeable education in a place like the Wayfinding Academy, a two-year institution of Portland, Oregon (https://wayfindingacademy.org/student-experience).

The Internet makes Illichian education more extensively possible in ways that were not there before its invention. One imagines for instance the Khan Academy (https://www.khanacademy.org/), which is not really an academy but rather a storehouse of academic "curricula," which appears to be tailored to replace or perhaps to supplement high school. (More teasingly perhaps is the Online University of the Left (http://ouleft.sp-mesolite.tilted.net/) with its promise of socialist online learning.) Now there is something genuine about the free-access-to-information and free-self-expression aspects of the Internet, and the optimistic anarchism of the Internet has inspired thinkers such as Richard Barbrook to write articles such as "Cyber-Communism: How the Americans are Superseding Capitalism in Cyberspace" (Barbrook, 2000), journalists such as Paul Mason to write books such as *Postcapitalism* (Mason, 2015), and science-fiction writers such as Cory Doctorow to write novels such as *Walkaway* (Doctorow, 2017) and *Down and Out in the Magic Kingdom* (Doctorow, 2003). More generally, Suoranta and Vaden (2010) *Wikiworld* suggests an "edutopia" as the site for "digital social creativity" (112) that actually changes society: "the still-beating heart of an interactive renaissance through the development of an information society is dependent on actual freedom in the sense of 'reconfiguring the co-ordinates of the possible'" (135). Essentially the Illichian Internet utopia depends not just on human goodwill but also upon motion—if it slows down it is in danger of disappearing. The seeming freedom of Internet information from authoritarian gatekeepers, the core of the fantasy life of the Internet, nonetheless inspires utopian dreams of freedom from capitalism in some writers.

But all of this alternative cyberspace utopianism, carrying the "learning web" to an actual Web and bringing it into a greater light as a set of utopian dreams, would need an additional social transformation if it were to focus upon learning about the Capitalocene, which would entail (1) utopian dreaming as an instrument of social change and (2) ecological knowledge of the Capitalocene (and of the precipice it leads the social world toward) as a basis for utopian dreaming. In its present state, the Internet "learning web" appears as a supplement to the degree-mongering, which serves as a purpose for much of what counts as knowledge today. Its practical knowledge, moreover, serves purposes which aren't prone to integration into an overall vision of ecotopia.

Thus, the Ivan Illich utopian dream might be compelling in some ways, but will not develop spontaneously out of capitalist society—some other prior social transformation, unrelated to schooling, would have to facilitate it. This is also true of the utopian dream suggested briefly in Peter McLaren's tour-de-force (McLaren, 1993/1986) ethnography *Schooling as a Ritual Performance*, a book with a unique yet entirely accurate perspective upon teachers and students as viewed through an anthropological lens and in context. McLaren's book is basically an ethnographic study of 7–8 grade students in Toronto, somewhat off-topic for the curriculum proposed here—but it suggests a unique lens through which student lives can be examined and through which social transformations can be conjured and brought to life, and so it is relevant here.

During Freire's lifetime, the real-life educational scholar and global activist Peter McLaren was an important friend and a co-creator of critical pedagogy. In *Schooling as a Ritual Performance*, McLaren proposes that the teacher performs the role of a "liminal servant," someone with the "best attributes of a teacher working within a liberatory pedagogy" (114). The liminal servant appears in the book as someone who can midwife praxis, though McLaren is not so explicit as to how this could happen; the ethnographic study conducted in *Schooling as a Ritual Performance* does not have to address the possibility of the liminal servant's success because there were no fully equipped liminal servants for McLaren to observe at the school where he did his observations.

For McLaren (1993/1986), the student is already part of collective determinations of ritual, albeit ones devoted to politically reactionary purposes given the absence of liminal servants at his school. *Schooling as a Ritual Performance*, with its analysis of structured formal education, ritual, performance, and Victor Turner (which McLaren used heavily in *Schooling*), is relevant to the idea of a curriculum for the Capitalocene because it brings into the discussion of everyday life the existence of a *ritual continuum*. The ritual continuum is the series of postures and gestures we perform in doing what we do every day. It can only be described with respect to ritual time, which is what McLaren does with his description of the school day in *Schooling as a Ritual Performance*. Under capitalism, congruently, the ritual time of the workday can be measured as separating out into socially necessary labor time and surplus labor-time, which is what Karl Marx suggested in *Capital*.

One important connection *Schooling* makes is that the ritual continuum is a trap both in terms of capitalist labor and repressive education, of the type McLaren studied before writing that book. Liminality, sometimes the door-opening prerequisite to a better life both in *Schooling* and in the anthropology of Victor Turner, disrupts the ritual continuum in favor of an alternate curriculum, which is what the curriculum for the Capitalocene will have to be if it is to disrupt the tidy function of college classes toward degree-mongering and career advancement.

In the ethnographic study of *Schooling as a Ritual Performance*, McLaren observed students doing math, bullying other students after school, attending mass, at a school dance. The key for McLaren was to observe ritual learning firsthand. He describes, for instance, a school mass:

> The schoolwide Christmas mass certainly did not appear to capture the attention – let alone the commitment – of a large number of students. In fact, this mass was one of the shallowest I have ever experienced; it utterly failed to engender feelings much different than the most perfunctory instructional rite.
>
> Student: what a bore, man. We might just as well have been doing math.
>
> A somber, highly structured occasion, with teachers 'scanning' the aisles and watching for students who were moving or talking, the mass became a forum of student resistance. Students laughed, joked, discussed the latest drug deals and bumped into each other. During the sign of peace, hands were squeezed hard; some students even tried to pull each other over the pews - all this despite the strict controlling forces of the teachers. (155).

These descriptions add meaning to thinking about teaching adults (which is what the curriculum for the Capitalocene will have to be) as observations of a ritual order;

there is a ritual order which rules adult lives as well as the lives of children . At any rate, at this Catholic school, in mass, like in math, the ritual order was both imposed and subverted in unimportant ways, rather than being consciously used to create learning experiences. The sum result is a situation in which, as McLaren points out, "power and privilege become 'somaticized' (168)." Ultimately, the repetition of such experiences leads the lower-class students at the school represented in *Schooling* ("St Ryan") to become obedient participants in a lower class. He writes, "It is no understatement to say that in many ways the classroom at St Ryan mirrored the workplace of the factory" (225). McLaren, then, used his study of Catholic school as an attempt to reveal the creation of social classes in everyday life.

The notion of a "class divide" in educational ritual practices is of course well documented; what immediately comes to mind is Paul Willis' (1977) *Learning to Labour*, an ethnographic study of British schooling. For a somewhat updated American context, Annette Lareau's (2011) *Unequal Childhoods* is probably the most thorough representative of educational literature documenting how the more privileged classes learn differently. More useful than either middle- or lower-class educational ritual experiences, both of which were documented by Lareau, would be liminal education, education that tries to break through the "crust of everyday meanings" and to establish learning as an experience both intellectual and somatic which can create a "politics of the possible" (McLaren 290).

The politics of the possible will have to be examined first before it can be created. Of course, the politics of the possible is going to be different for the college student than what it was for the 7–8 grade Toronto school that was the subject of McLaren's ethnographic study. For schoolchildren, the politics of the possible is one in which a good education is made possible or one in which it isn't. For the college student, the politics of the possible is one in which a hint might be planted in the students' minds that learning need not all be about career advancement.

There is nonetheless a politics of the possible for the college student, and it should be examined in light of the typical college student's entry or re-entry into life in the so-called real world as a functioning adult. Can the Capitalocene last indefinitely? Is the Capitalocene to end tragically, with capitalism collapsing amidst the ruination of career aspirations and mass homelessness and starvation? Or are there alternatives? What can the functioning adult do, with her options of citizenship and collective labor and human solidarity? Perhaps it is realistic to imagine one course as something merely to incite the student search for answers, rather than expecting it to do everything.

In *Schooling*, the liminal servant is just one of three archetypes of teacher role-playing; the other archetypes, common to the lives of legal minors in lower class North America but not to be desired, are the "teacher-as-entertainer" and the "teacher-as-hegemonic-overlord" (116). These other two roles were pre-Internet adaptations to classroom management in lower-class settings. The "teacher-as-entertainer" was, for instance, a show-biz alternative to the "pale, monochromatic" (222) landscape of the classrooms McLaren studied, and the "teacher-as-hegemonic-overlord" was, as McLaren said, a "conditioned reflex of the culture's consensus ideology."

The teacher-as-liminal-servant role suggests an idealized teacher in a subversive way. As McLaren (1993/1986) puts it, the liminal servant is "the bringer of culture and is ever-cognizant of her shamanic roots" (115). The liminal servant is an emcee of scholastic occasions of release from the social structure: a teacher, but also a real-time liberator. The "liminal" state, for McLaren as for Victor Turner (who popularized the term), is the sensation of being in a ritual state in which the status system of a society is suspended, and in which the ritual participants are open to participation in a state of communitas (Turner 106–107), a state of unstructured communion, which suggests the immediate experience of utopia.

In Turner's terms, the liminal servant is a "ritual elder" (Turner 96). McLaren's liminal servant is still by profession a teacher, but more importantly, she represents the magic of learning to be made real through performance. Opposed to the efforts of the liminal servant in McLaren's study is the "ritual system" (81), which governed the life of the school McLaren studied and which appears typical of schools that legal minors are required to attend. The liminal servant's opposition to the "ritual system" is a teacherly one—as McLaren says, "To a greater extent than the other pedagogical types, the liminal servant is able to help students crack the prevailing cultural crust and discover alternative meanings" (117). With the liminal servant, then, one hopes for preliminary gestures toward ultimate social liberation.

The key concept of *Schooling as a Ritual Performance* linking education with the ritual system is that of *ritual knowledge*. McLaren defines the concept and explains its general application in the school he studied as follows:

> Broadly speaking, ritual knowledge can be conceived as interaction between the form and content of curriculum and the lived culture of the students. The streetcorner state offered a more vital form of ritual knowledge since. It permitted closer correspondences between the social body and the physical body. The mechanical formality, regimentation and invariance which were part of the instructional rites to a large degree stripped the classroom cultural landscape of its organic symbols. (204).

As McLaren describes it, the imagined success of the liminal servant is placed in opposition to teaching-as-it-often-is, a mere conformism of the everyday ritual of life in school. McLaren's conclusion about his ethnographic site for *Schooling as a Ritual Performance*, essentially a Catholic middle school, was that "it is no understatement to say that in many ways the classroom at St Ryan mirrored the workplace of the factory; there was a distinct isomorphism between the use of space and time in the school and the daily itinerary of the factory worker" (225). So, there is little hope for McLaren's school.

Toward the end of *Schooling as a Ritual Performance* McLaren hints at an ultimate end of praxis, in social revolution:

> Schools must begin to give measurable shape to our dreams for a more just society by becoming not only laboratories for critique, but also strongholds for purposeful and life-giving symbols. The realm of school ritual must illuminate these symbols, fire them in the crucible of wisdom, and shape them on the anvil of liberation. Only then will we move confidently towards the brink of the fearful abyss that Victor Turner calls the antistructure, grab hold of that rope of snarled symbols, and propel ourselves to the other side where knowledge and freedom meet once and forever. (256).

Scholastic praxis, then, is to make of school a laboratory for social critique and a place where the utopian dream can be revealed. The genuinely transformative sites for praxis, the rituals of student liberation, will be those established in a revolution, in which the liminal state can be achieved most productively.

It is more or less up to the reader to decide what will count as a ritual of student liberation, invoking the ritual knowledge necessary for genuine change. The usual ritual is the ceremony of college graduation, after which the student becomes deinstitutionalized; but college graduation is a way of justifying college on its proclaimed merits, which may not suit the purposes of praxis. The political campaign, the strike, the protest, and the demonstration all might count as credible rituals, though they take place outside of formal educational frameworks. (What comes to mind is the storming of the Bastille, the pivotal event of the French Revolution.) There will also be a problem of how to make the liminal ritual an educational one; for every Bastille storming, there should be a *Declaration of the Rights of Man and the Citizen*.

Another candidate for the liminal event, given the topic of environmental crisis which shadows this essay, would be the natural disaster. In this regard, the reflections of Rebecca Solnit's (2010) *A Paradise Built in Hell* might be of relevance. Solnit discusses, with copious reference to historical examples, instance of human behavior in disasters, human-caused and otherwise, in which strong feelings of solidarity impel people to help each other when in ordinary existence they might have very little to do with each other. Perhaps the future disasters of the Capitalocene can provoke revolution.

The point of discussing liminal events, then, is that people are otherwise hemmed in by a ritual continuum, which is the clock- and calendar-time manifestation of social structure. In the summary chapter for *Schooling*, McLaren argues for a basic congruence between the drudgery of school life for the students he observed, and the drudgery of labor-time in capitalist life. The qualities that fill time with drudgery, their basic components, are micro-measurement of the temporal units and external control of ritual time. Outside of all that, during a social transformation, a liminal "anti-structure" in the classroom can free up time for a social anti-structure, and liberate society for participation in historical time in what Turner called a "social drama." Praxis can begin in the classroom, but it must take place as part of an overall change in the larger society, of historic, global, and environmental dimensions.

The curriculum proposed below will entertain such thoughts, encouraging a revival of utopian dreaming and a preparation for the social drama that, with luck, will put an end to the Capitalocene and usher in a better era of natural history.

A Curriculum for Utopian Dreaming and Praxis within the Capitalocene

The model selected here is that of a 10-week college class in the setting of a North American university, though it can be adapted to other circumstances. The great quantity of material suggested by this curriculum would seem to indicate a graduate class—the model as a whole would have to be greatly simplified if it is to be adapted to less "schooled" audiences and audiences with less time to spare for school activities.

Freire (1993) and Schugurensky (2014), there are three stages to the literacy method: (1) research, (2) thematization, and (3) problematization. A Freirean curriculum spread out over 10 weeks, then, might consider an adaptation of such thinking as follows: (1) Weeks 1–3 would explore the students' backgrounds, (2) Weeks 4–6 would explore the students' utopian dreams as against the background of the Capitalocene. (3) Weeks 7–10 would elaborate upon questions of action.

The curriculum proposed here starts with these guidelines:

1. The shape of your curriculum is going to depend upon your students. How are they situated in the "real world"? Thus, the first portion of the class should open a discussion about students and about how they are situated in the world. This real world can include historical contexts, most notably the environmental one in which the Capitalocene appears, but also that of the Age of Utopian Dreaming, the age in which utopian dreams informed popular notions of the future through concepts such as "progress."
2. One's course's electronic component should be open-ended. The "Illichian" aspect of the Internet is its promise of increased access to information and to society, and not the ease through which coders and their rule-prescribing administrators are able to transpose pre-Internet curricula onto institutionally mandated online courses.
3. In the section on action, collective, revolutionary options can be opened up as topics for discussion. Misiaszek (2018:30) sets up the problem as follows:

> Problem-posing in ecopedagogical spaces includes dialogue on how and why teaching socio-environmental issues is often systematically ignored, as well as the goal of constructing possible solutions, and actions toward those solutions, inside and *outside* of current economic, social, and political systems. There are two important aspects to such dialogue. First is that "outside" the current systems mean that ecopedagogical discussions are not limited to societies' current social structures and normative ideologies. Second is that transformation of the current systems themselves for increased socio-environmental justice is also part of ecopedagogical dialogue. These are both key tenets of Freirean Pedagogy.

In addition, the idea of ecopedagogy as citizenship education (see Misiaszek, 2016, 2018) is fine as one example among many (one thinks of Kahn's (2010) suggestions of Marcusean education and of traditional ecological knowledge) of an education that provokes students to think about what Misiaszek calls "socio-environmental utopias" (Misiaszek, 2018:109). If another world is to be possible, we have to be able to dream of the other world we want.

1. The standard trinity of race, class, and gender that one sees in college-level consciousness raising can be broached through the concept of the utopian dream. What would a world look like in which the social problems named by race, class, and gender had been solved? How could we go from the utopian dream of their solution to making such a dream possible?

2. The "generative themes" portion of the curriculum can be handled using the techniques discussed in Ira Shor's article "Monday Morning Fever: Critical Literacy and the Generative Theme of 'Work'" (104–121 of *Freire for the Classroom: A Sourcebook for Liberatory Teaching* (ed. Ira Shor)).

Four things need to "be there" to if the Freirean teacher is to foreground praxis, and empower students to change the world. They are as follows:

1. A potential (at least) social movement to support the utopian dream to be explored in the Freirean classroom—thus to attract student support for praxis.
2. An institutional basis for the Freirean teacher as a professional (i.e., for the teacher fantasy as historically/socially constructed beforehand).
3. A curriculum sufficient to prepare future generations for planetary healing.
4. An opening for the Freirean teacher to be a "liminal servant".

If not all of this is immediately available, some of it may have to "come later."

Here is a suggested curriculum for requirement #3 above, suited to a college-level 10-week class:

The first week will be dedicated to an exploration of students' existence in the world. Writing assignments will focus upon the construction of profiles.

Generative themes: career, society, location

The second week will have students locating specific utopian dreams within their existence. Students should be able to identify the desires for optimism in their lives. Writing assignments will focus upon how students see optimistic futures and their place within such futures.

Generative themes: progress, development

Week Three will explore the Age of Utopian Dreaming, with readings of history in that age. Students will be able to identify the various utopian dreams that occurred in various eras of history, against the background of capitalist history. Writing assignments will focus upon periods of history, which excite student interest.

Generative themes: history, utopian dreaming

Weeks Four and Five will have students meditating upon the concept of "the Capitalocene." Students are to find and report upon topics in environmental history. The conquest of the Americas, the "Great Acceleration," and the history of capitalism will all be explored. Research connecting personal experience to environmental history will be conducted.

Generative themes: bundled natures, climate change

Week Six will ask students what it would take to develop a utopia of sustainability, so as to avoid the dystopian implications of Week Five. How can global society prepare for the worst?

Generative theme: sustainability, dystopia

Weeks Seven and Eight will engage students in reflections upon possible actions they might take to apply the concepts learned in previous weeks. Writing assignments will reveal student choices in this regard.

Generative theme: the profile

Weeks Nine and Ten will be for discussion and student presentation.

Possible Accompanying Texts:

Texts in agroecology, permaculture, and the "Transition Town" movement, short or long depending upon the time available to read them, might be meaningful and useful in promoting the notion, depending upon where and how the class is situated in academic space. The point of such texts would be to show that ecotopia does not need to be imagined as some far-off fantasy having nothing to do with practical reality. Rather ecotopia is something that can be incorporated into one's life here and now, within limits.

Students might also do well to read Ursula K. LeGuin's *The Dispossessed* to deal with the social prerequisites and limitations of the sort of utopian society that would be significantly different from the one in which we currently live. Such a society would be democratically organized for collective survival, unlike with the current society, which is organized for the domination of "value" over people's lives (and thus the constant need to pander after good jobs, money, food, shelter, and other accoutrements of survival).

Hilke Kuhlmann's *Living Walden Two* looks at what a small group of people did to live a utopian dream, specifically the Twin Oaks commune in Virginia, connecting it to the success these people had and the challenges they faced and currently face.

Edward Goldsmith et al.'s *Blueprint for Survival,* Ernest Callenbach's *Ecotopia,* Sally Miller Gearhart's *The Wanderground,* and Marge Piercy's *Woman on the Edge of Time* are all ecotopian fictions from the 1970s, and all appropriate texts for discussion of how ecotopia is to be imagined 50+ years after they were written. An additional conversation-topic surrounding these novels would have to do with their oldness—why is ecotopia something which was for the most part laid out in the 1970s? Was the path to ecotopia lost after that period of history, and, if so, how might we rediscover it?

Cory Doctorow's *Walkaway* is a present-day story of a computerized dystopia, which becomes a utopia. It would be an especially useful accompanying text for an online version of a utopian-dreaming course.

References

Barbrook, R. (2000). Cyber-communism: How the Americans are superseding capitalism in cyberspace. *Science as Culture, 9*, 5–40. Retrieved June 27, 2018, from http://www.imaginaryfutures.net/cybercommunism_art.pdf

Bonneuil, C., & Fressoz, J.-B. (2015). *The shock of the anthropocene* (D. Fernbach, Trans.). London: Verso.

Dawson, A. (2016). *Extinction: A radical history*. New York, NY: OR Books.

Doctorow, C. (2003). *Down and out in the magic kingdom*. New York, NY: Tor.

Doctorow, C. (2017). *Walkaway*. New York, NY: Tor.

Freire, P. (1993). *Pedagogy of the oppressed* (20th anniversary edition; M. B. Ramos, Trans.). New York, NY: Continuum.

Freire, P. (2000/1970). *Cultural action for freedom*. Boston, MA: Harvard Educational Review.

Freire, P. (2004). *Pedagogy of indignation*. Boulder, CO: Paradigm Publishers.

Gadotti, M., & Torres, C. A. (2009). *Paulo Freire: Education for development*. Hoboken, NJ: Wiley Online Library. https://doi.org/10.1111/j.1467-7660.2009.01606.x
Grollios, G.. (2016). *Paulo Freire and the curriculum* (N. Gakoudi, Trans.). New York, NY: Routledge.
Illich, I. (1970a). *Deschooling society*. New York, NY: Harper & Row.
Illich, I. (1970b). *Celebration of awareness: A call for institutional revolution*. New York, NY: Pantheon.
Illich, I. (1973). *Tools for conviviality*. New York, NY: Harper & Row.
Kahn, R. (2010). *Critical pedagogy, ecoliteracy, and planetary crisis*. New York, NY: Peter Lang.
Mason, P. (2015). *Postcapitalism*. New York, NY: Farrar, Straus, and Giroux.
McLaren, P. (1993/1986). *Schooling as a ritual performance* (2nd ed.). Lanham, MD: Rowman and Littlefield.
McNeill, J. R., & Engelke, P. (2014). *The great acceleration: An environmental history of the anthropocene since 1945*. Cambridge, MA: Harvard University Press.
Misiaszek, G. W. (2018). *Critical global citizenship education*. London: Routledge.
Pinar, W. F. (2013). *Curriculum studies in the United States*. New York, NY: Palgrave Macmillan.
Schugurensky, D. (2014). *Paulo Freire*. London: Bloomsbury.
Solnit, R. (2010). *A paradise built in hell: The extraordinary communities that arise in disaster*. London: Penguin.
Suronata, J., & Valden, T. (2010). *Wikiworld*. London: Pluto.
Taylor, P. (1993). *The texts of Paulo Freire*. Philadelphia, PA: Open University.
Turner, V. (1969). *The ritual process*. Ithaca, NY: Cornell UP.
Willis, P. (1977). *Learning to labour: How working class kids get working class jobs*. New York, NY: Columbia University Press.

Chapter 3
Critical Environmental Education with an Ecosocialist Vision

Constantine D. Skordoulis

Introduction

The growing awareness of ecological crisis and the ecology movement developing since the early 1960s have represented a vigorous critique on the idea that the actually existing social and economic order is capable of guaranteeing continuous "progress for all," and that harnessing nature is inherently positive and that all problems pertaining to it could be solved.

Up against this challenge, there have been attempts at updating environmental policy on a global scale. The first, known throughout the world, was the Club of Rome report (Meadows et al., 1972). This report documented the environmental problems and put forth an international policy against demographic growth, wastage of natural resources, environmental destruction, etc. This study and others following were a double-edged sword. On the one hand, the international governmental administration retook the initiative on environmental questions and undertook a discussion on the prognosis and the solutions to be put forth. On the other, these studies shored up pessimistic views on the world's future and were a further impetus to the ecology movement. The world economy's existing order lost its aura of superiority; its finality and its mechanisms were questioned from within. At the same time, these analyses led to lists of demands that tended to promote world planning and a political regulation of the economy. In fact, they came into direct conflict with the market economy, economic liberalism, and government deregulation offensives on the agenda throughout the world at the time.

No later than the mid-1980s, a second offensive on the environmental field proved necessary especially in terms of concrete policy, to these contradictions. The Bruntland report (Our Common Future) adopted by the UN General Assembly in

C. D. Skordoulis (✉)
School of Education, National & Kapodistrian University of Athens, Athens, Greece
e-mail: kskordul@primedu.uoa.gr

A. S. Gkiolmas, C. D. Skordoulis (eds.), *Towards Critical Environmental Education*, Critical Studies of Education 14,
https://doi.org/10.1007/978-3-030-50609-4_3

1988 was an expression of this. It is entirely marked by the conviction that although the existing economic order unfortunately harms the environment, it is also in a position to make the necessary corrections. Thus, it claimed to bring together the elements for a more balanced form of economic growth (what was later termed "sustainable development").

The 1990s saw a deepening of the contradiction between promises of new international regulations and this very system's destructive social and environmental impact. The Rio Declaration, which came out of the Earth Summit (1992), certainly set forth certain principles, such as the precautionary principle, which did represent progress in awareness about the elements of the ecological crisis. Neither Agenda 21 nor the international conventions on biodiversity and climate change have led to the solutions needed.

With the birth of the WTO further subjecting the environment to the effects of liberalized international trade, these conventions have had very little effect. Proclamations in defense of biodiversity are powerless against ongoing damage to the natural environment. On the political level, they run up against the interests of agrochemical and pharmaceutical multinationals that seek to take hold of living organisms by increased use of GMOs and patenting genomes.

The USA Administration, backed by energy company lobbyists, rejected the Kyoto Protocol (1997) on the greenhouse effect. The fragile agreement that the other industrialized countries reached in 2001 does not put any onus on rich countries to implement domestic measures cutting their gas emissions and amounts to giving up the already inadequate targets included in the initial protocol. In fact, the protocol proposes a target of only a 5.2% reduction in CO_2 emissions, despite the fact that these emissions would have to be reduced by more than 50 percent in order to keep CO_2 levels in the air under 550 ppm—double the level of the pre-industrial era—and stop global average temperatures from rising more than 2 °C.

More than 120 billion dollars over 10 years had been announced in Rio for environmental protection policies on the world scale. In 1996, only 315 million dollars had been invested. Between the ideas put forth by the Bruntland report, and again in Rio, and the dominant ultra-liberal economic model, the latter has won the day for the time being.

The September 2002 Johannesburg summit ended in resounding failure: no significant international measure was adopted. The summit showed by contrast that the big multinationals had succeeded in a few years in imposing their views within the international institutions. They have thus acquired a tribune to promote privatization of public resources and goods through notions like "public–private partnership."

NGOs, with a strong presence at Rio and since then, have sometimes let themselves get caught up by the environmentalist discourse of the G7 and international institutions. In the future, they will have few other choices from a complete integration as an ecological veneer on capitalism or to go back to a radical ecological criticism, which for many of them means going back to their roots.

All of these elements explain how the line of the American administration was outflanked at the time of the Bali conference (December 2007), and that this conference led to a relative unblocking of the negotiations in view of a new international treaty, which is supposed to take over from the Kyoto protocol.

The "road map" adopted in Bali referred in a precise way to the quantified conclusions that should be drawn from the 2007 report of the IPCC. The ink was not yet dry on this document when the G8 decided in favor of a reduction of global emissions by 50% in 2050, without mentioning either the upper end of the range of global reduction put forward by the IPCC (85%), or the reduction objective concerning the developed countries (from 80% to 95% between now and 2050), or the intermediate reduction objectives for these countries (from 25% to 40% between now and 2020), or the decrease in global emissions from 2015.

At the beginning of 2008, the European Commission proposed to the member states and to the European Parliament an "energy-climate package" (20 percent of reduction in emissions, 20% of gains in energy efficiency and 20% of renewable energy—including 10% of biofuels in transport, between now and 2020). This "package" is lower than the recommendations of the IPCC and incompatible with the objective adopted by the Council in March 1996, of a maximum rise of 2 °C.

In autumn 2008, in the context of the "financial crisis" unleashed by the economic recession, several member states (Italy, Poland and the Czech Republic in particular) and industrial sectors (automobile, iron and steel) contested the contents and especially the methods of the "package." The Council of December 2008 maintained the symbolic formula 20–20–20 but, essentially, it is now nothing but a facade. The employers have very largely obtained satisfaction on two key points: exemption from payment for emission rights for sectors "exposed to international competition" and for the coal-fired power stations of the new member states, and the massive externalization of efforts toward the developing countries, by means of the CDM (nearly 70% of reductions in emissions could be delocalized to the South).

In the context of globalization, a vast offensive is underway to impose a system of "marketing the right to pollute" on the world level in order to reduce the quantity of greenhouse gases. Advocated by the United States, this mechanism was accepted by the European Union. This is a dangerous development that must be reversed.

First, it opens the way to strengthening under-developed countries' dependency on the North. In a mechanism assigning each country an exchangeable pollution quota, the decision-making power belongs to those who hold financial power to trade in pollution as they see fit. The highly indebted countries of the South and the East would run the risk of selling their quota to the Northern countries, though the latter pollute the most by far. In fact, the system aims to make pollution a commodity, hence a source of profit.

Finally, it must be emphasized that the purpose of this mechanism, the key element of the neoliberal offensive in the environmental field, is to defuse the subversive power of the ecological critique, which raises a challenge to the overall functioning of the capitalist system. It aims at restoring credibility to the idea that the market is the best instrument in the fight against pollution, that more capitalism would make for intrinsically "cleaner" capitalism.

Environmental Education as a Critical Process

Recent developments in the field of Environmental Education cannot be viewed separately from the neoliberal system's offensive against the environment.

The idea that the market is the best instrument in the fight against pollution had to be introduced in Environmental Education, which traditionally has been a field of education fostering radical critique of societal and economic practices and policies.

Environmental Education has been recognized as a very important policy component by the Tbilisi Declaration (UNESCO, 1978), and this has been reassured by Agenda 21 (UNCED, 1992).

The Tbilisi Declaration taking into account that "the environment is a system" and that "the human environment has structure, operation and history," suggests that countries should "develop the systemic approach in analysis and management of physical and human ecosystems" (UNESCO, 1978, p. 20). The systemic approach in Environmental Education constitutes an ideal approach for the teaching of ecological ecosystems, as complex self-organized systems. In addition, the systemic approach inevitably takes into account the social, economic, and political dimension, as factors that could potentially harm the global ecosystem.

Environmental Education is a radical process with deeply political implications aiming at the fostering of an active citizen, who is able to resist and to take action for the protection of the environment.

This dimension of Environmental Education requires to precisely identify "who takes the decisions" in social life. There is no critical dimension in environmental issues without reference to the social hierarchy and to its decision-making procedures. This kind of education requires both a radical critique of value systems and resistance accompanied by concrete decisions and practices. Thus, criticality comprises deep knowledge; choice; accountability; search for truth and justice. Students need to continuously "sharpen" their critical mode of thinking through the study of complex environmental issues, as they encounter competing aspects and values, which have to criticize and ultimately to take position.

Recently, certain alterations were introduced to the original character of Environmental Education by the Decade of Education for Sustainable Development (2005–2014) program (UNESCO, 2005).

Over the past 30 years, Environmental Education is being affected not only by the overall restructuring of the educational system globally according to the OECD guidelines but also by a policy that envisages to make it more complicit with the general environmental policy of the world's rich as outlined in the previous section.

A critical point in the transformation of Environmental Education was the introduction of the notion of "sustainable development" as a major international educational priority which took place at the Rio conference in 1992 with the Article 36 of Agenda 21. Although Education for Sustainable Development appeared, at first, as a component within Environmental Education, a differentiation was soon made evident, which presented Environmental Education as a singular field of reduced importance in comparison with the new Education for Sustainable Development agenda (Kahn, 2008).

Despite the fact that sometimes Environmental Education and Education for Sustainable Development are mentioned as identical fields, there are conflicting views as to the one that has to be developed.

On the one hand, there are scholars who favor Education for Sustainable Development, considering this approach as more holistic compared to the traditional Environmental Education, which mainly focuses on the understanding of ecological relationships. On the other hand, there are educators who find the Education for Sustainable Development framework to be limited in theoretical scope (Kahn, 2008), adapting to the existing dominant ideas and concepts and failing to address the profound challenge of the global environmental crisis as we know it (Tsoubaris & Georgopoulos, 2013).

The interconnection between Education for Sustainable Development and the green development agenda is associated with the European Union Green Public Procurement (GPP) directive and supports the EU's efforts to become a more "resource-efficient economy." The fact that green procurement has gained such an important place compared to other environmental policy issues is indicative of the tendency to promote what Kahn (2010) refers to as "green consumerism," i.e., reducing environmental awareness to making individual consumer choices of "greener product."

This type of consumerist culture amounts to a public pedagogy (Giroux, 2005) where the values of individualism, competition, and profit turn into a dominant monologue to any social question posed. Furthermore, according to Kahn (2010) "the culture of green consumerism can easily serve as a way to socially reproduce the dominant social order through a wide variety of narcoleptic shopping exercises that profess our collective salvation via the growth of individuals' mounting credit debt."

What has also been highlighted is the instrumentalist nature of Education for Sustainable Development, as a form of education which "delivers and propagates experts' ideas about sustainable development" (Tsoubaris & Georgopoulos, 2013). It has been noted that Education for Sustainable Development is based on the belief that future technologies and the market will solve environmental problems while addressing current economic, political, and social effects as "natural," therefore considering their transformation as "unnatural" (Fassbinder, Misiaszek, & Thordarson, 2010).

Kahn (2008) has suggested that Education for Sustainable Development consists part of a top-down agenda for business-as-usual, which is driven by green development goals rather than schools' local educational and environmental framework.

Taking into account both the institutional factors, as well as its educational context and structure, Education for Sustainable Development is closely linked to economic and political aspects of the green development agenda.

This critique of Education for Sustainable Development has opened the way for educators in Europe and the USA to search for an alternative ethical, educational, political, and pedagogical paradigm focusing on new concepts such as "ecopedagogy" (Kahn, 2008), "eco-justice pedagogy" (Bowers, 2001), "critical ecopedagogy" (McLaren & Houston, 2004, 2005), and "revolutionary critical ecopedagogy" (McLaren, 2013).

Critical Pedagogy in its various forms such as "radical pedagogy," "liberatory pedagogy," "revolutionary pedagogy," "oppositional pedagogy," and "border pedagogy" (Au, 2007; De Lissovoy & McLaren, 2006; Giroux, 2003; Hill, 2003) aims to provide a means by which the oppressed may begin to reflect more deeply upon their socioeconomic conditions and take action to improve their life. At the same time, Critical Pedagogy highlights the ideological/moral, the collective/social, the subjective/context-driven, and the praxis (reflective action) dimensions of this undertaking.

Paolo Freire recognized that ecology has gained tremendous importance at the end of the 20th century. It had to be present in any educational practice of a radical, critical, and liberating nature (Freire & Shor, 1987).

Recent scholarship (McLaren & Houston, 2004, 2005) in critical pedagogy emphasize that "greening" critical pedagogy ought not diminish its radical intent or its goal of transforming oppressive social and economic conditions. Drawing on the field of political ecology (McLaren & Houston, 2004, 2005) argue for critical revolutionary pedagogy to be informed by a dialectics of justice. The two sides of this dialectic being *environmental justice*—the question of the unequal distribution of harmful environments between people—and *ecological justice*—the justice of the relationship between humans and the rest of nature (McLaren & Houston, 2005:169).

McLaren and Houston (2004, 2005) aim to "map out what a dialectics of environmental and ecological justice might look like for critical and revolutionary educators by examining how justice toward those exploited under the capitalist class system is increasingly shaped by environmental concerns" (ibid., pp. 169–70). Fundamentally, what this means for students is a drawing out of the complex web of relations between (1) local, place-based environmental injustices; (2) historical injustices arising out of the circuits of capitalist social, political, and ultimately economic relations (including racism and colonialism); (3) the impacts of industrial and neoliberal processes on the planet's ecosystems; and (4) the ideological production of nature under capitalism. The exploration of this web of relations operates to allow student teachers to develop critique of educational and social practice at Allman's different levels of truth. Understanding the material and ideological production of nature (and indeed environmental crisis) as a social and historical process highlights how our ideas of what matters in nature is never fixed, uniform, or stable. What an ecosocialist project for education as McLaren and Houston (2004, 2005) suggest "broadly illuminates is precisely how the present state of nature is neither inevitable nor desirable – and that ecologically and socially just alternatives exist" (ibid. p. 173).

In this chapter, I will present the alternative based on the theoretical premises of Marxist ecology, what has been termed "ecosocialism," and I will describe its potential to educate students to the underlying causes of environmental destruction, which is nothing else than the capitalist relations of production and the thirst for profit.

The Ecosocialist Perspective: Its Philosophy and Politics

The ecology movement's fundamental achievement, which has brought about an in-depth change in public awareness of environmental questions, has been and remains the understanding of the extent to which late capitalism has destroyed the environment. Destruction of nature has reached a point that imperils all humanity. Here, as in the case of a world nuclear war, it is a question of survival. However, contrary to the danger of nuclear destruction, it is a question that is always "new" and becoming increasingly serious and obviously so. Nonetheless, the fact that the movement views the environmental question as vital to all humanity does not justify doing what most ecologists do: seeking out interclass solutions, eliding the importance of class struggle against capital. The distinction between those who have an interest at maintaining the system, whatever the cost, and those who have an interest in abolishing it has not been erased; quite the contrary.

Ecosocialism is both a vision for a society based on ecological and social justice and a philosophy advocating the democratic organization of this society based on the free association of producers. It differs from other radical discourses, because it has a more anthropocentric character and thus avoids the rejection of humanism as presented, for example, in deep ecology.

In the last 30 years, in an attempt to provide a coherent analysis of the combined ecological and capitalist crises, ecosocialist scholars have debated notions such as "the conditions of production" (James O'Connor), "ecological materialism" (John Bellamy Foster), and "productivism" (Michael Lowy) among others (Skordoulis, 2019). These theoretical insights have provided a deeper understanding of the causes of the environmental crisis and exemplified that an entirely different relationship of humankind to the environment is an urgent necessity. This new relationship, based on a caring model for both humans and the environment, will not be simply the result of individual changes in behavior. Rather it needs a structural change in the relationships among humans and in the mode of production of social existence.

Marx and Ecology

Ecologists have sometimes accused Marx and Marxists of an uncritical attitude toward industrial society and the damage it does to the environment. Marxists need to "break radically," according to Michel Löwy, "with the ideology of linear progress and with the technological and economic paradigm of modern industrial civilization" (Löwy, 2005, 16). Although there are some on the left, who would still agree with Löwy, it would seem that the belief that Marx offered a powerful and coherent approach to ecology has been gaining ground in recent decades. John Bellamy Foster, a prominent defender of a Marxist inspired ecology, claims that: "Few involved in ecosocialist discussions today doubt the importance of Marx's foundational contribution to the ecological critique of capitalism" (Foster & Burkett, 2016).

The question as to why this aspect of Marx's historical materialism was either not known or forgotten for so long has a number of answers. The development of Marx's thinking on ecology occurred in the last two decades of his life and many of his writings and notes from that time are still not published. The orthodoxy of Second International Marxism and the technological optimism of the early Russian revolution and the dogmatism and industrial strategy of the later Soviet Union were also a factor. The Frankfurt School and other Western Marxist writers, very influential after the Second World War, were mainly interested in culture and aesthetics and rejected the idea that the Marxist dialectic could be applied to nature (Fagan, 2019).

Although there were socialist ecologists in the 1960s and 1970s, it wasn't until the 1980s, with the work of Marxists such as Ted Benton and Elmar Altvater, and the foundation of the journal "Capitalism, Nature, Socialism," that the idea of a Marxist ecology began to take shape.

The charge of "productivism" against Marx was challenged with greater theoretical rigor by John Bellamy Foster's "Marx's Ecology" (Foster, 2000) and Paul Burkett's "Marx and Nature" (Burkett, 1999). Both argued, in these and many publications since, that Marx's analysis of capitalism was an ecological one. In Marx's theory, the drive for profit and the accumulation of capital was based on an unlimited appropriation of natural resources which have a natural limit.

More recently, Kohei Saito's "Karl Marx's Ecosocialism" sets out to demonstrate "the immanent systemic character of Marx's ecology, that there is a clear continuity with his critique of political economy" (Saito, 2017, 12). What gives added interest to Saito's claims is his use of what then were the still unpublished ecological notebooks of Marx from 1865 to 1868 (Fagan, 2019).

Research carried out over the last two decades has demonstrated that there was a powerful ecological perspective in classical Marxism. Just as a transformation of the human relation to the earth was, in Marx's view, an essential presupposition for the transition from feudalism to capitalism, so the rational regulation of the relation to nature was understood as an essential presupposition for the transition from capitalism to socialism.

Marx and Engels wrote extensively about ecological problems arising from capitalism and class society in general, and the need to transcend these under socialism.

This included discussions of the nineteenth-century soil crisis, which led Marx to develop his theory of metabolic rift between nature and society. Basing his analysis on the work of the German chemist Justus von Liebig, he pointed to the fact that soil nutrients (nitrogen, phosphorus, and potassium) were removed from the soil and shipped hundreds and thousands of miles to the cities where they ended up polluting the water and the air and contributing to the poor health of the workers. This break in the necessary metabolic cycle between nature and society demanded for Marx nothing less than the "restoration" of ecological sustainability for the sake of "successive generations" (Marx, 1976a: 636–639, Marx, 1976b: 754, 911, 948–949).

In line with this, Marx and Engels raised the main ecological problems of human society: the division of town and country, soil depletion, industrial pollution, urban maldevelopment, the decline in health and crippling of workers, bad nutrition, toxicity, enclosures, rural poverty and isolation, deforestation, human-generated floods, desertification, water shortages, regional climate change, the exhaustion of natural resources (including coal), conservation of energy, entropy, the need to recycle the waste products of industry, the interconnection between species and their environments, historically conditioned problems of overpopulation, the causes of famine, and the issue of the rational employment of science and technology.

This ecological understanding arose from a deep materialist conception of nature that was an essential part of Marx's underlying vision:

> Man lives from nature, i.e. nature is his body, and he must maintain a continuing dialogue with it if he is not to die. To say that man's physical and mental life is linked to nature simply means that nature is linked to itself, for man is a part of nature (Marx, 1974, 328).

Not only did Marx declare in direct opposition to capitalism that no individual owned the earth, he also argued that no nation or people owned the earth; that it belonged to successive generations and should be cared for in accordance with the principle of good household management (Marx, 1976b, 911).

The Political Economy of Ecosocialism

Ecosocialism is the current in the workers' and ecology movements most sensitive to the interests of workers and peoples of the South. It breaks with the productivist ideology of progress and opposes the infinite expansion of an environmentally destructive mode of production and consumption. It understands that "sustainable development" is impossible within the framework of the capitalist market economy.

Ecosocialists call for a genuine change in culture. This is impossible without an in-depth technological reorientation, seeking the replacement of current energy sources with other, nonpolluting and renewable ones, such as solar energy. This means the first issue at hand is the question of control over the means of production, and above all over decisions relating to technological change (Skordoulis, 2019).

An overall reorganization of the mode of production and consumption is needed, based on criteria foreign to the capitalist market: people's real needs and environmental protection.

This means an economy in transition to socialism, based on the peoples' own democratic choices of priorities and not the "laws of the market." This would be a planned economy, able to find lasting ways of overcoming the tensions between satisfying social needs and ecological imperatives. It would be a transition leading to an alternative way of life, a new culture beyond the reign of money, consumer habits, and the endless production of environmentally harmful goods.

The mode of capitalist production comes into fundamental contradiction with nature and the natural evolution process. For capital, only the quantitative aspect is decisive, determining the relation between labor time and money in the framework of the law of value; qualitative and global relations cannot be taken into consideration.

The spread of capitalist commodity production cannot respect preexisting modes of social organization. Occupying the space needed for a smooth production process, energy supply, and distribution must go ahead without taking the natural environment into account.

It is not capitalism's lack of wisdom that brings about environmental destruction, but the very logic underlying the system.

Capitalist rationality determines the movement of individual capital. However, competition among capital makes the system as a whole irrational.

Nowadays, commodity production governs all sectors of social life, while the social process of production has become more and more fragmented. Property relations have become more and more centralized; competition among owners of the means of production keeps them from becoming entirely frozen.

This has led to the same major ecological problems in all advanced capitalist countries. Here is yet more proof that these problems cannot be viewed as "breakdowns" or "system failures"; they correspond to this system's logic (Skordoulis, 2019).

Capitalist production also reshapes its own consumers. Thus, individuals' behavior is a factor aggravating the ecological crisis and hampering a solution to it. A flagrant example of this is what could be called the "dictatorship of the car," i.e., the ecologically catastrophic system of private cars, promoted by automotive industry marketing, by bourgeois individualist ideology, by the deliberate dilapidation of public transport, but also by the urban structure of major cities, which forces workers to commute long distances. However, individual changes in behavior can only exert a minimal influence on the fundamental environmentally destructive nature of capitalist production.

Progress and Crisis

The main achievement of the ecosocialist movement is the way it questions the concept of "progress." It has demonstrated that we can no longer speak as during the beginning of capitalist development of a positive development of the productive forces, only restricted by private ownership of the means of production or developed at the expense of the proletariat. More and more, capitalism, having survived much longer than historically necessary for the development of the productive forces, is transforming productive forces into destructive ones. But this also means that these forces cannot be liberated as such, that is, used in a socialist system on behalf of all.

Competition among capitalists leads to periodical overproduction crises, revealing that a considerable quantity of energy and materials has been invested in commodities that don't sell. Furthermore, the market promotes the production of superfluous products in use-value terms (advertising, various drugs, arms, etc.) but with an exchange value that makes for big profits.

Competition and the race for profits are the ultimate reason behind non-respect for environmental regulations, use of toxic substances, inadequate quality testing, falsifying content listings, unauthorized dumping of waste, etc. (Skordoulis, 2019).

Instead of being a source of social progress, the development of productivity has led to an intensification of exploitation of labor power, production choices disconnected from social and ecological imperatives, and chronic overproduction crises.

Due to of the impact of capitalist production on the environment, destruction of the natural basis for human societies has reached a new level.

The ecological crisis is worldwide and, in the competitive context inherent to capitalism, can only be viewed only as a common evil; certain causes of the ecological crisis go back many years, others are the products of the combined development of various separate factors.

For this reason, it is difficult to establish and date their temporal and physiological causes. In the same vein, mastering the ecological crisis calls for time and investments that would be the undoing of all bourgeois concepts of input/output cycles.

Finally, contrary to what is observed in classical economic crises, in capitalism's harmful social consequences and even in the aftermath of military conflicts, dominated and exploited classes can only be made to foot part of the bill for ecological crisis. However, it is undeniable that oppressed classes bear the brunt of the burden, especially in dependent countries. This is truer still, given the interaction between social and economic crisis and ecological crisis.

Ecosocialists understand that market and profit logic are incompatible with environmental protection. While criticizing the lack of pro-environmental politics in the traditional currents of the workers' movement, they understand that workers and their organizations are an essential force for transforming the system.

Critical Environmental Education with an Ecosocialist Vision in Practice

Ecosocialist politics make the connection between opposition to the destruction of the environment and neoliberal policies—indeed, to the capitalist system itself—and an education system which feeds, supports and reproduces both the production and consumption sides of an unsustainable economic system.

McLaren and Houston (2004, 2005) argue that critical and revolutionary educational praxis is increasingly shaped by and through ecological politics and they analyze schooling as one site of environmental injustice before embarking on a broader discussion of how justice toward nature more generally may be substantively linked to the objectives of critical revolutionary pedagogy.

This has been further exemplified in the work of Hill and Boxley (2007) who set out a series of progressive egalitarian policy principles and proposals that constitute a democratic Marxist and ecosocialist manifesto for schooling and teacher education for economic and social justice. This is based on a democratic Marxist theoretical framework and on a structuralist neo-Marxist analysis. They also draw out a set of strategic connectivities between these programmatic ends and forms of resistance to neoliberal hegemony to be found in the interstices of contemporary educational systems. In this paper, Hill and Boxley (2007) call for a transformative change throughout teacher education, throughout schooling and education, by cultural workers within the apparatuses of education and the media, and throughout multiple layers of civil society. They suggest a series of specified sets of principles and a program for critical and socialist educators, through which to engage with the Radical Right. The programmatic matters they outline in their paper represent the potential for a galvinizing union of current environmental and social concerns and labor rights and conditions battles in the educational sphere, and suggest a project of united action against the capitalist vision of education in the twenty-first century.

More recently, Chattopadhyay, Gahman, and Watson (2019) edited a special issue of the journal "Capitalism Nature Socialism" under the title: "Ecosocialist Pedagogies." In their "Introduction," they explicitly state that neoliberal economic empires operate through control, domination, and accumulation at the expense of species at the bottom of the hierarchies they institute. One way to resist this is to consistently engage with radical projects and simultaneously refine teaching techniques responding to increasing socioecological injustices. They suggest that an ecosocialist approach to education is distinct from other radical and libertarian approaches, and this position can yield interesting insights into philosophies of education.

They also argue that radical ecosocialist pedagogical frameworks can promote nonhierarchical student cooperation on projects in class and in local communities introducing theories that resonate with the everyday experiences of students from oppressed groups and can sanction participatory and action-oriented learning, encourage problem-based learning, and a dialogical method of problem-solving, as well as promote experiential learning by following "other" histories and context-specific struggles.

Adopting the aspects mentioned above, we have designed an Environmental Education course in the Department of Primary Education at the National and Kapodistrian University of Athens placing emphasis on the actual philosophical debates and conflicting policies in the environmental movement with the purpose of developing students' criticality and understanding that environmental problems are in the last analysis social and political problems and that their permanent solution is inextricably linked with change both on the level of culture and of social structure. We have been teaching an Environmental Education course for the last fifteen years in the traditional way (Mandrikas, 2012; Mandrikas, Assimopoulos, & Skordoulis, 2015) and throughout this period aspects of ecosocialist critique have been gradually introduced in the course. Namely, more than one third of the course evolved around Environmental Philosophy and assessment of the global environmental problems. It is this part of the course that I am discussing in the following paragraphs.

The course consists of two parts: an introduction to the main currents in environmental philosophy including ecosocialism and an examination of the main global environmental problems focusing on energy and climate change.

In the first part of the course, students are initially introduced to philosophical concepts such as anthropocentrism, ecocentrism, conservation, and preservation. This is accomplished by a historical introduction to the main ideas and policies of the nineteenth-century environmentalists such as John Muir, Gifford Pincheon, and Henry David Thoreau, followed by a historical account of the emergence of Environmental Ethics as a distinct branch of Applied Philosophy after the pioneering work of Arne Naess and his co-thinkers. In this first part, the development of the radical philosophies of Social Ecology by Murray Bookchin, Ecofeminism, and Ecosocialism are discussed in detail along with their critique of Deep Ecology.

Drawing from the principles of Critical Pedagogy, our teaching method is related to the dialogical processes in the classroom, which have their roots in the variety of cultural communities in the classroom, characterized by their social and environmental living conditions. So, we have been constantly trying to keep the traditional deductive lecture system to a minimum and instead to promote the dialoguing process in the classroom. After a short introduction, students are encouraged to discuss and debate the differences and tensions between these radical approaches to environmental philosophy.

Questions introduced by the lecturer evolve around two axes: Nature-human/culture relation (*How are humans related to nature? Are we different and separate, full a part of it, or what? What is the relationship between human culture (language, literature, human artifacts, technology* etc.*) and nature?*) and the value of nature (*Does nature have intrinsic value? Is value primarily in individuals or in "wholes?"*) of humans (*What is the relative value of humans in relation to nature?*) and of animals (*how ethical is to experiment on animals? has life an inherent value? what is the social meaning of life*?).

In the second part of the course, topics for discussion are mainly focused on the global ecological crisis and its main manifestation, climate change.

Jensen (2002) has developed a useful framework that categorizes knowledge that informs and promotes pro-environmental behavior and other forms of action in terms of four dimensions: (1) knowledge about the environmental issue/problem—principally, its nature and extent, and the scientific and technological knowledge relevant to it; (2) knowledge about the underlying social, political, and economic structures, and how they contribute to creating particular environmental problems; (3) knowledge about how to bring about changes in society through direct or indirect pro-environmental behavior; (4) knowledge about the likely outcome or direction of possible changes (Hodson, 2009, p. 8).

In the second part of the course, we have complied with these four dimensions in order to organize methodologically the debates among the various student groups in the classroom.

Topics for discussion are selected from the international ecosocialist literature and discussed in this course were the spiral of poverty on the planet (*why the 1% of the population of the earth owes the 50% of the global wealth?*), nuclear energy (*is nuclear energy the clean energy of the future?*), fossil fuels, and oil drilling (*what is the culture behind the use of fossil fuels? the use of the car and the culture of individual transportation? the environmental and geopolitical consequences of drilling for oil in the south east Mediterranean Sea? the current crisis in Iraq, Syria and the Middle East?*), megacities and public space (*what is the logic behind urban planning in megacities? why there is a lack of public space in poor neighborhoods? what is the significance of social movements like "reclaim the streets"?*).

Students are given material (newspapers, TV broadcasts, scientific reports, etc.) with conflicting accounts of the causes and the societal impact of the different environmental issues and are encouraged to debate and discuss in groups. These discussions improve their argumentation skills and their critical thinking approach.

In a preliminary assessment of the course, we have observed that a substantial number of students through their discussions gradually came to understand that the underlying cause behind the main environmental problems is social antagonism and in the last analysis an antagonism between social classes on a global scale.

In discussing the impact of the course with the students, some declared that it had an immediate effect on their everyday life attitudes, such as water saving, energy saving, recycling, and reduction of consumption. Some others declared that it had an intermediate effect, such as a desire for more information, a desire for action ("something has to be done"), development of an interest about urban parks, resistance against repression, etc.

Other students said that they began to collect articles from newspapers and journals for future use in the class, exactly as they have seen during the course.

It is our opinion that these findings show that students' judgments concerning the course they had attended were favorable. The critical approach seems to be well experienced, as long as students disentangle its characteristics and evaluate them positively. So, it is our belief that Environmental Education can be developed around an ecosocialist vision based on the appropriate teaching methodology.

References

Au, W. (2007). Epistemology of the oppressed: The dialectics of Paulo Freire's theory of knowledge. *Journal for Critical Education Policy Studies, 5*, 2. Retrieved from http://www.jceps.com/?pageIDOarticle&articleIDO100

Bowers, C. A. (2001). *Educating for eco-justice and community*. Athens: The University of Georgia Press.

Burkett, P. (1999). *Marx and nature, a red and green perspective*. New York, NY: St Martin's Press.

Chattopadhyay, S., Gahman, L., & Watson, J. (2019). Ecosocialist pedagogies: Introduction. *Capitalism Nature Socialism, 30*(1), 26–30. https://doi.org/10.1080/10455752.2019.1587241

De Lissovoy, N., & McLaren, P. (2006). Ghosts in the procedure: Notes on teaching and subjectivity in a new era. In K. Cooper & R. White (Eds.), *The practical critical educator: Critical inquiry and educational practice* (pp. 151–165). Dordrecht: Springer.

Fagan, G. (2019). *Was Marx an ecologist? Reviews/Karl Marx, Friedrich Engels, Exzerpte und Notizen, Marx-Engels-Gesamtausgabe, IV, 18, article 6182* (p. 1294). Berlin: deGruyter. Retrieved from http://www.internationalviewpoint.org/spip.php?

Fassbinder, S., Misiaszek, G., & Thordarson, J. (2010). Critical pedagogy, ecoliteracy and planetary crisis: The ecopedagogy movement: A book review symposium. *Journal for Critical Education Policy Studies, 8*(1), 461–482.

Foster, J. B. (2000). *Marx's ecology*. New York, NY: Monthly Review Press.

Foster, J. B., & Burkett, P. (2016). *Marx and the earth: An anti-critique* (Historical materialism book series) (Vol. 115). Boston, MA: Brill. https://doi.org/10.1163/9789004288799

Freire, P., & Shor, I. (1987). *Pedagogy for liberation: Dialogues on transforming education*. London: Macmillan.

Giroux, H. A. (2003). Critical theory and educational practice. In A. Darder, M. Baltodano, & R. D. Torres (Eds.), *The critical pedagogy reader* (pp. 27–56). London: Routledge Falmer.

Giroux, H. (2005). Cultural studies in dark times: Public pedagogy and the challenge of neoliberalism. *Fast Capitalism, 1*(2), 75–56. Retrieved from http://www.uta.edu/huma/agger/fastcapitalism/1_2/giroux.html

Hill, D. (2003). Global neo-liberalism, the deformation of education and resistance. *Journal for Critical Education Policy Studies, 1*, 1. Retrieved from http://www.jceps.com/?pageID'0article&articleID'07

Hill, D., & Boxley, S. (2007). Critical teacher education for economic, environmental and social justice: An ecosocialist manifesto. *Journal for Critical Education Policy Studies, 5*(2), 28–77. Retrieved from http://www.jceps.com/index.php?pageID=article&articleID=96

Hodson, D. (2009). Putting your money where your mouth is: Towards an action-oriented science curriculum. *Journal for Activist Science & Technology Education, 1*(1), 1–15.

Jensen, B. B. (2002). Knowledge, action and pro-environmental behaviour. *Environmental Education Research, 8*(3), 325–334.

Kahn, R. (2008). From education for sustainable development to ecopedagogy: Sustaining capitalism or sustaining life? *Green Theory and Praxis: The Journal of Ecopedagogy, 4*(1), 1. Retrieved from http://www.academia.edu/163841/From_Education_for_Sustainable_Development_to_Ecopedagogy_Sustaining_Capitalism_or_Sustaining_Life

Kahn, R. (2010). Producing crisis: Green consumerism as an ecopedagogical issue. In J. Sandlin & P. McLaren (Eds.), *Critical pedagogies of consumption*. London: Routledge. Retrieved from https://www.academia.edu/172713/Producing_Crisis_Green_Consumerism_as_an_Ecopedagogical_Issue

Löwy, M. (2005). What is ecosocialism? *Capitalism Nature Socialism, 16*(2), 15–24.

Mandrikas, A. (2012). Towards a critical environmental education for pre-service elementary teachers. *Kritiki: Science and Education, 12*, 29–44. (in Greek) Retrieved from http://www.hpdst.gr/publications/article

Mandrikas, A., Assimopoulos, S., & Skordoulis, C. (2015). In-service teachers' views towards an environmental science education course designed around inquiry principles. *Journal of Studies in Education, 5*(1), 108–138.

Marx, K. (1974). *Early writings* (p. 328). New York: Vintage.

Marx, K. (1976a). *Capital* (Vol. 1, pp. 636–639). New York, NY: Vintage.

Marx, K. (1976b). *Capital* (Vol. 3, pp. 911–948–949). New York, NY: Vintage.

McLaren, P. (2013). Seeds of resistance: Towards a revolutionary critical ecopedagogy. *Socialist Studies, 9*(1), 84–108. https://doi.org/10.18740/S4QG6G

McLaren, P., & Houston, D. (2004). Revolutionary ecologies: Ecosocialism and critical pedagogy. *Educational Studies, 36*(1), 27–45. https://doi.org/10.1207/s15326993es3601_4

McLaren, P., & Houston, D. (2005). Revolutionary ecologies: Ecosocialism and critical pedagogy. In P. McLaren (Ed.), *Capitalists & conquerors: A critical pedagogy against empire*. Oxford: Rowman & Littlefield.

Meadows D. H., Meadows D. L., Randres, J. & Behrens III W. W. (1972): *The Limits to Growth: A Report for the Club of Rome's Project on the Predicament of Mankind*, New York: Universe Books.

Saito, K. (2017). Karl Marx's Ecosocialism. In *Capital, nature and the unfinished critique of political economy*. New York, NY: Monthly Review Press.

Skordoulis, C. (2019). *From environmental ethics to political ecology: An ecosocialist perspective*. Athens: Propobos Publishers. (in Greek).

Tsoubaris, D., & Georgopoulos, A. (2013). Gauging the potential of socially critical environmental education (EE): Examining local environmental problems through children's perspective. *Journal for Critical Education Policy Studies, 11*(4), 132–158.

UNCED (United Nations Conference on Environment and Development). (1992). *Agenda 21*. Rio de Janeiro (3–14 June), Brazil.

UNESCO. (1978). *The final report: International conference on environmental education*, Tbilisi (14–26 October), ED/MD/49, Paris.

UNESCO. (2005). *UN decade of education for sustainable development 2005–2014, draft international implementation scheme*, Paris, ED/2005/PI/H/1.

Chapter 4
This Branch Is an E: Conversations About a Curriculum for Earthlings

Jackie Seidel and Stephanie Westlund

Through the telling of three stories, followed by conversations from our positions as an aunt/scholar and a mother/scholar, this chapter works to articulate and understand the grief we feel at observing children become violently schooled in ways that disembody and separate them from what David Abram (1997) called 'the animate earth' (p. x) and from the places that languages and cultures emerge(d). The global community of human and non-human others faces an unprecedented ecological crisis characterized by the mass extinction of species and cultures and an unparalleled flow of human beings across borders. In this context, schools could be a radical nexus point in societies where diverse humans of diverse ages and cultures collaborate on living well in this world in this time, in kinship with one another and with other species. We envision the creation and fostering of pedagogies and curriculum that embrace our interdependencies across species, place and time so that our

J. Seidel (✉)
Werklund School of Education, University of Calgary, Calgary, AB, Canada
e-mail: jlseidel@ucalgary.ca

S. Westlund
Earthling and Independent Scholar, Calgary, AB, Canada

A. S. Gkiolmas, C. D. Skordoulis (eds.), *Towards Critical Environmental Education*, Critical Studies of Education 14,
https://doi.org/10.1007/978-3-030-50609-4_4

relationships are continuously remembered and cared for. In this light, critical environmental education becomes an interpretive pedagogical stance that can shape life in classrooms and schools towards more healthful and just ways of being and learning together with the earth and all our diverse kin.

Inspired by the invitation for chapters for this book, we engaged in focused conversation around stories of shared interest and concern that involved Stephanie's children (Jackie's nephew and niece). We drew on duoethnography as a dialogical method (Norris, 2008; Norris, Sawyer, & Lund, 2012; Sawyer & Norris, 2013; Seidel & Hill, 2015) to ground ourselves and explore our individual and shared responses to these stories. We spoke on the phone, made notes on random papers, emailed back and forth, and met to combine and edit these into this chapter. These stories and conversations happened over many months in 2017 and 2018 in Calgary, Alberta, Canada, on the traditional territories of the people of the Treaty 7 region in Southern Alberta, which includes the Blackfoot Confederacy (comprising the Siksika, Piikani and Kainai First Nations), the Tsuut'ina First Nation and the Stoney Nakoda (including the Chiniki, Bearspaw and Wesley First Nations). The City of Calgary is also home to Métis Nation of Alberta, Region III (see https://www.ucalgary.ca/indigenous-strategy/resources/territorial-acknowledgements). The Bow River is the main water source for the city as well as many communities downstream. It also provides crop irrigation for much of arid southern Alberta. Originating high in the Rocky Mountains, the river is fed by snow and glacier melt. Climate change is provoking serious concern about current and future water security for the entire region.

Our conversation illuminates some of the philosophical, practical, political and cultural challenges in overcoming culturally entrenched 'traditions' that remain prevalent in curriculum, schools and schooling. The conversational methodologies of storytelling and duoethnography serve as radical ruptures in the patriarchal and colonial forms of inquiry inherited from Western Science. As ecological and relational methods, they enable us to enter the inhabited 'small' spaces of the lives of young children.

Story 1: This Branch Is an E!

Stephanie and Jackie:

Ellen[1] *is three and a half. The half makes a difference. Like so many humans before her, for at least several tens of thousands of years, she's discovered that she can make marks that mean something, marks that others can read and interpret. In this case, she's learned that an E stands for her name: Ellen. She's excited by this new knowledge. Her small hand grips her pencil tightly, and she marks E after E after E on papers she scatters around the house. The direction of the letter doesn't*

[1] The children's names have been changed.

matter to her yet, nor does the number of lines coming out… 3… 4…5. They are all E's. This is her name. She has marked the page. She's thrilled and proud. She tapes her pages to the wall for others to see. She gifts Jackie with a drawing of a stick person with a large head and E's all over the page.

We go outside to play in the yard. It's a crisp day in early fall. The leaves have yellowed and are falling to the ground. Suddenly Ellen runs excitedly around the corner of the house, shouting and carrying a huge poplar tree branch that is bigger than herself and declares, 'This branch is an E!!!'

Jackie:

This moment was so exciting to me! The branch was clearly an E. I shared this moment with the students in my undergraduate English Language Arts methodology course that I was teaching at the time. What could we say about this? That Ellen loves E's because that is her name. To her, her name is inscribed in the E on the paper (which comes from trees), drawn with her own hand, but also in nature, in a tree branch. For a 3-year-old, there is not yet any separation between who she is in the world and how she is in the world. *Everything is still whole*. Her body, herself, her world… same thing. I asked my students when and how they thought school would teach this out of her. I was reminded of Derrick Jensen's book *Walking on Water* (Jensen, 2004), which is an extended meditation on the crushing impacts of industrial education on human possibility(s) and creative potential. He reflects on the etymology of 'education': '… from the Latin root *e-ducere*, meaning "to lead forth" or "to draw out." Originally, it was a midwife's term meaning "to be present at the birth of"' (p. 15). He reflects on the ways our inherited industrial schooling system, with its explicit intention to standardize humans for work in the capitalist economy, is engaged in 'routine dehumanization' (p. 215). Elsewhere, I have called attention to how our approach to language instruction in schools, particularly when reduced to discussions of 'literacy' success and failure as linked to a nation's economic competitiveness, is connected not only to dehumanization but also to contemporary (and historical) processes of the mass extinction of biocultural diversity: that is, both the terminal loss of species of all kinds and also of human cultures and languages (Seidel, 2007). Language extinction's connection to ecosystem and biodiversity loss is exceptionally well documented in research such as by Nettle and Romaine (2000).

Stephanie:

I worry, too, about school teaching this out of Ellen—sometimes, I am reluctant to enrol her in kindergarten next year. We were very deliberate in choosing the Reggio-inspired playschool she currently attends several mornings per week—some days the children and teachers spend the whole time outdoors, while other days they play in a bright, window-full classroom with natural objects and plants. These days, her pockets are *always* filled with sticks and rocks and plant matter. Her knees have grass stains. One day, she came home and proudly announced that she peed in a bush. She tells me about investigating and taking apart an old wasp's nest. In the winter, the children slide down the hill on their bums in the snow. They make snow forts. They play outside until they are too cold to move. In all seasons, they look for signs of wildlife. They make mudpies. They hide in bushes. They pick dandelions.

On inside days, if they are being too loud or wiggly (they share a building with quiet office workers), the teachers take them outside to blow off steam—to run around. They are taught to listen to—and to *read*—their bodies and make connections with the world around them.

Jackie:

Similar to how you describe Ellen's preschool education, a critical environmental education would be committed to holding the ecological connections and relations between language and place and life together. It would hold them 'in place' and 'in time'. It would be committed to caring for all the branches, in both a metaphorical and literal sense, that our actual living depends upon. This seems like important healing work in Western capitalist culture that depends on separations and divisions to maintain its power structures. These structures continue, in the name of 'progress', to perpetrate violent forms of colonization and racism, gender violence, as well as violence against nature. They depend on us never acknowledging our kinship with other living beings across time and place. These structures are rapidly destroying the planet's ability to sustain life. The breath of the world is *in that branch*, indeed, it *is* that branch, in a very profound way. Ellen's declaration serves for me as a powerful reminder of this. Her relationship with the branch is the same as her relationship with what she wrote on the page.

Story 2: The Died Deer

Jackie and Stephanie:

We walk together by the Bow River. It's a warm cool May day. Sweater weather. We've been here many times before. The destination is the playground on the other side of the river, in a park about 2 km away. While crossing the pedestrian bridge, we are distracted by a large rock on the gravelly river bank. It seems incongruous and we wonder why we haven't noticed it before. Curious now, instead of continuing to the playground, we go down to the rocky beach to investigate. It's a dead deer, stomach bloated and legs chewed and bones exposed by a predator or scavenger, most likely a coyote. The children are fascinated. Stephanie calls the City to report it. Not long after, two young men arrive in a small garbage truck. They study the deer and are clearly uncomfortable with the situation. A small crowd gathers on the beach and on the bridge above. Eventually, the young men make a decision. Laughing nervously, they pull on blue latex gloves and wrap a cable around the deer's body and secure it with a large hook. One of them flips a switch and suddenly a winch rapidly drags the deer's body towards the truck. It crashes violently against the back bumper of the vehicle and remains stuck there for a moment, the winch motor whirring and pulling. Suddenly, the deer's body flips and disappears into the back of the truck. Another switch is operated and the back of the truck closes. The young men drive away.

On the slow walk home, snacks in hand, we see other living creatures. Baby Canada Geese (ubiquitous in Calgary in the spring) waddling around after their mothers, swimming in a long line in the river... and an eagle flying overhead. We stop to watch them. Camden talks avidly about the deer. He wonders what might happen to its body. We don't know. He suggests that they should shoot it into outer space. He doesn't want it to go to the garbage dump. He imagines how he could build a Lego model of the scene, including the garbage truck with a working winch. Ellen, nearly 3 years old, is quiet and curled up in the stroller. Stephanie stops, picks Ellen up and asks what's wrong. Ellen starts to cry and bursts forth with, 'The died deer's name was Frieda!!!' For the rest of the walk, being safely carried and held, Ellen is silent except for telling Stephanie that she is going to 'tell daddy about the died deer'. Upon arriving home, she runs into the house excitedly calling for her dad. He crouches down and listens to her story. Ellen's spirits lift and she runs away happily to play.

Jackie:

I've been fascinated by Ellen naming the 'died deer', as well as by her overwhelming need to share this story with her dad... Also, that both children were so affected by what they witnessed that day. Camden processed the experience through talking about it all the way home, imagining how he could remember it through recreating the scene (with his hands! See our discussion at end of chapter about Kelly Lambert's research), and Ellen needed to share the story with a significant person who hadn't been there, her father. Ben Okri (1998) calls us 'storytelling beings' or 'homo fabula'. I wonder if storytelling is important to critical environmental education, and if so, how? In my teacher education work, I visit many schools and rarely see storytelling built into children's days in any significant way. Their own stories and experiences and questions (like Camden wanting to know what would happen to the deer's body after the garbage truck) are not seen as important to formally build into the education process. I am concerned about any environmental education that would simply be more of the same, more talking down to children, or talking at children, rather than engaging them in significant and embodied experiences where there would be a story to tell later on. Imagine if your child came home and told you all kinds of amazing experiences of what happened to his body while playing outside, or what he witnessed (or didn't witness) on a classroom walk through the community...

Stephanie:

It's true! Lorri Neilsen (1998) reminds us that stories grounded in real lives have the power to move people from being spectators or consumers to being actors (p. 266). But perhaps, the most important aspect of storytelling and embodied experiences is that they require imagination! We can empathize with other beings, both creatures and humans, and imagine ourselves in their place. As Hannah Arendt (1968) observed, 'the more people's [and creatures'] standpoints I have present in my mind while I am pondering a given issue, and the better I can imagine how I would think and feel if I were in their place, the stronger will be my capacity for representative thinking' (p. 241). Imagination really is crucial to understanding the complex relational interdependencies that form the basis for all life. It strikes me that an important part of critical environmental education involves the capacity to imagine ourselves in a web of relationships with all living and non-living beings—to recall our mutual relationships and the cycle of life.

Jackie:

I remember when you brought Ellen over last fall to watch the honey bees in my yard cull the drones from the colony. As we stood observing them, she commented that she wanted to see the 'died bees', so she remembered Frieda, the died deer, even 6 months later. It was really something significant for her. Learning about death and life cycles, human and non-human, seems like something important missing from our education system. Critical environmental education would probably have something to do with learning about death, about the cycles of living and dying that sustain this world.

Stephanie:

You might remember that our pet fish (who lived with us for more than 4 years) died this year during the frigid, snowy month of February. The kids wanted to bury him outside, but instead, I put him in the freezer with a promise to bury him in the garden when the ground thawed. We only buried him recently—and I was truly struck by how sad the children were all over again. It was as though he had just died. They told stories about him all day. Giving him the proper burial and storying it was an important step in mourning their pet.

Story 3: Windowless Classrooms and Recycling Guilt

Stephanie:

A few weeks ago, Camden came home from school and announced, 'I petted a bumble bee today at recess!' 'You petted a bumble bee?' I repeated, trying to make sure I'd heard him correctly. And he excitedly told me about deciding to pet the bee during his lunch break and how it was soft and it didn't even notice him. My son, the bee whisperer.

This image of Camden petting a bee—an experience of profound contact with another living creature—stands in stark contrast to the classroom at the local public school into which he goes every day: a bleak, windowless room lit by artificial light. There's nothing alive in there—except the 50 children and two teachers, of course. Camden told me, after his first few days in that room, 'We never know what it's like outside. We don't know if it's sunny, or cloudy, or raining'. There's no one to blame, except maybe the original architect (and I can't help but wonder what he/she thought about children and the kind of spaces they deserve!). Over the years, different experts have been consulted about renovation possibilities, even simply adding skylights or tubes into the ceiling to bring in natural light. Can't. The asbestos in the ceiling cannot be disturbed.

In school, Camden is taught about recycling. He does not allow me to send anything disposable in his lunch. No plastic bags. No juice boxes. No individually wrapped granola bars. Anything sent in a disposable bag comes home uneaten. He won't even take it out of his lunch bag for fear that he will be shamed for bringing something disposable.

About three times per year, families are invited to the school for a special lunch or picnic and we are also asked to bring our own plates and cutlery (nothing to throw in the garbage please). Camden knows how to compost, too. Each class has a compost bin where the children can put banana peels and other food waste as well as things like paper towels. Last year, his teachers also had a worm composter in the classroom. Sometimes, Camden got to feed the worms his kiwi skins.

At least there's recess. Oh wait. His lunchroom supervisor has a habit of keeping the children in. They're being too loud or too wiggly, she complains. She waits until they are quiet and still before sending them out. This usually takes a long time. Sometimes, the school bell brings them back inside only a few minutes after they go out. I complain to the principal and teacher about this, but they tell me that 'group consequences' are important for improving student behaviour.

Next year I hope for windows.

Stephanie:

The thought of 50 tiny bodies in a classroom with no windows makes me feel sick.

Jackie:

I once refused to teach in a classroom with no windows, and argued to the principal that caged animals in a zoo are treated more ethically and that people would be marching in the street if 50 animals were locked up in a small space without natural light or air for so many hours at a time. What does this say about how we think about children? That we value their lives and days even less than those of caged zoo animals? We are animals—*mammals*—co-evolved and co-emerged in ecological relation with the rest of life. Indeed, here in Calgary in the long northern winter, children and teachers have very little access to natural light during those months. It seems even more important that much time and importance would be committed to taking our animal bodies outside.

I am concerned about the 'disciplinary' reasons given for not allowing the children to go outside at Camden's school. This is also connected to industrial schooling and the capitalist economy—both have a vested interest in disciplining the animal body. It happens in schools and it happens in factories (and zoos, too!). There have recently been many news stories about how Amazon treats its employees, even denying them bathroom breaks (and, therefore, we can assume also breaks to go outside in fresh air, breaks to eat nourishing food, time to rehydrate). There is always another body for the neoliberal economy to gobble up. It doesn't care what someone's name is, or what their story or experience is, as long as they are a disciplined, standardized worker (and well prepared for that role by schools!).

Stephanie:

My own academic background has led me to be familiar with the research about the importance of nature contact—for example, research shows that the symptoms of children with ADHD are often exacerbated in windowless rooms while natural spaces alleviate them (Taylor, Kuo, & Sullivan, 2001), and blood pressure and feelings of anger and aggression tend to rise in windowless spaces (Hartig, Evans,

Jamner, Davis, & Gärling, 2003). Electroencephalography (EEG) studies have found that participants in green spaces 'showed lower frustration, engagement, and arousal, and higher meditation' (Nieuwenhuijsen, Khreis, Triguero-Mas, Gascon, & Dadvan, 2017). A recent Stanford University study revealed that a 90-min nature walk resulted in both reduced levels of self-reported rumination and reduced neural activity in the subgenual prefrontal cortex—an area of the brain linked with mental illness (Bratman, Hamilton, Hahn, Daily, & Gross, 2015). A Canadian study drawing on data from almost 1.3 million people in 30 cities revealed the significant protective effect of living near trees—living in residential areas with more trees reduced Canadians' risk of an early death from several common causes by 8–12% (Crouse et al., 2017). Meanwhile, Max Planck Institute researchers used magnetic resonance imaging (MRI) to demonstrate that participants living closer to a forest generally had healthier amygdala structures (Kühn et al., 2017).

Despite being familiar with this research, every day I send Camden to that place without windows. I did suggest to his teacher that it might be possible to bring nature *into* the classroom—to use grow lights to grow seeds and plants and experiment with what can happen, even in a windowless room. She seemed excited by the idea, but I guess she was too busy to follow this through.

Jackie:

As a teacher and teacher educator, I imagine that this is the case. The workload and performance demands on contemporary teachers are incredibly high. They have no time to think. Also, they have been inculcated themselves into the same system so that many truly believe their job is to prepare children to be future workers. This enables 'progress-oriented' education to act as if children's present lives, or what is going on in the world outside of the school walls, is of no consequence. Time and place, and everything that matters in the present, are completely displaced. Growing plants is an extra frill that takes time away from literacy and math drills. Capitalism has no interest in people actually thinking. Busyness is good for business.

It strikes me that Camden's experience of 'environmental education' is making him feel shame and anxiety and guilt. It may be intended to 'make' him care about the world, but instead, it has become about rule following and being an obedient good boy. However, Camden is already deeply in love with the world. He petted a bumble bee! He knows how to garden, that food grows from seeds, and that it requires hard work and patience, certain weather, that one hail storm can destroy it. I wish he were learning this at school, as well as from you. There are many children at his school who do not have these outdoor, life-nurturing experiences that you intentionally create and share with your children.

Stephanie:

True! Camden's experience at school is dislocated from his place. Actually, you've reminded me about the work of scholar and peacebuilder John Paul Lederach (2005) and a critique he received from a Mayan priest. After Lederach gave a presentation about his 'integrated peacebuilding framework' to a group of Guatemalans, an indigenous Mayan priest came up to him. While 'your framework captures many things', the priest said, '[i]t is missing one overarching element' (p. 140). Lederach couldn't imagine which 'political, economic, or historic piece' he had missed. The priest continued:

> Your framework is missing the earth and skies, the winds and rocks. It does not say where you are located. In a traditional Mayan view, if there is a problem in the community, the first thing we would ask is: Did you greet the sun today? Did you thank the earth for the corn? It is not the only thing, but it is the first. We always must know where, [in] what place and time, we are located. (p. 140)

This is what is missing in Camden's school—the earth and skies, the winds and rocks. They rarely acknowledge where they are located. They can't even see outside to know if it is sunny or raining!

Jackie:

Rather than the guilt-inducing and limited environmental education that he's getting, I would prefer an ecological education that fosters a deep sense of connection, kinship and responsibility both in and across times and places. They should know intimately what the weather is! We have barely begun to imagine what such an education might look or feel like. It would be an abrupt break from how things are organized now, and it's hard to envision how that might even happen, given that any small curriculum change (particularly in Mathematics or Language Arts/Literacy) seems to result in parental protest and zero changes. This demonstrates how tight those strangling bonds of industrial/capitalist schooling really are.

Stephanie:

Yes! Environmental education in school should not *just* be about recycling and composting, about understanding and managing the natural environment, about so-called sustainable development. Rather it should be actively working to articulate and understand that *we are nature.* As the current research into microplastic pollution around the world is demonstrating there is no separation: what we do to the world, we do to ourselves.

As a parent, I want school to go beyond recycling and composting, to teach my children that *THEY ARE NATURE.* And as part of nature, their bodies send them (and those of us around them) messages. Rather than keeping the children in from recess when they are 'misbehaving', I want school to acknowledge and teach them that when they have that wiggly feeling or feel like shouting, their body is telling them to move, to go outside, to change their environment. I want school to stop teaching children to ignore these bodily impulses but rather to notice and listen to them. What environment brings forth these different feelings? What type of movement (or lack thereof) brings forth the wiggles and what brings a sense of calm? This, to me, would be the most useful form of environmental education—education that teaches children that their animal body is *always* interacting with everyone and everything around them.

Jackie:

This is really the point that Catherine Keller (1986, 1996, 2002), my favourite ecofeminist scholar, makes in her astute analysis. She carefully traces the intertwined historical developments and patterns of patriarchy, colonialism and capitalism, and how invested they all were and are in maintaining a sense of separation and disembodiment to maintain their power. Of course, schools and schooling are completely bound up in this history, as they emerged from and alongside it. This is why schools are so hard to change, and why little tweaks make no difference to their overall structures. Any kind of schooling that is not bound up in these historical and present structures is seen as 'alternative' or fringe, or as a threat to 'tradition'. We see this in teacher education at the university too. In our undergrad program, we don't have any courses that explicitly have anything to do with 'environmental education', much less courses that explicitly teach about the kinds of connections we are thinking about here (body, history, interspecies relations). This coming year I will be teaching four different undergrad courses, in essentially four different disciplines. I feel fragmented and exhausted in advance. How must the students feel? Sometimes, it seems teacher education serves mostly to reinforce the way things have been in the past, including a vision of human life that is violent and destructive to both the human and more-than-human communities. This conversation is challenging me to think about how I might bring some of these ideas, especially about bodies and embodiment, into those courses in new ways.

Stephanie:

I am reminded of a cold winter day several months ago when Ellen had been cooped up at home and she was being impulsive and uncooperative. It was −15 °C outside, but her body was sending the message that she needed a change in her environment. We put on our warmest clothes and took our sleds to the local park. We were the only people in the park, and we walked up and slid down the small hill many times. About 15 minutes into our adventure, Ellen sat in my lap, looked into my eyes and declared (without prompting), 'I feel calmer now'. I nodded and replied, 'Oh that's good. I feel calmer too. You seemed like you were feeling a little silly before'. And she said, 'Yes I was really getting crazy. And now I'm calm'.

I learned an important lesson that day: that a 3-year-old child can know, understand *and* intellectualize the messages her body is sending her. She can recognize that a change in her environment and moving her body makes her feel different after only a few minutes. This is what we can and should be teaching the children in our lives. And this shouldn't just be a home project. I want schools to do this too!

Jackie:

This reminds me of neuroscientist Kelly Lambert's research, which you introduced me to. It's a constant frustration to me that the profound implications of research like this, which should be applied to environments like schools or adult work situations, are largely ignored.

Stephanie:

One particular study has important implications for schools and our lives with young children as well as environmental education. In her neuroscience laboratory, Lambert and her students studied and compared rats whose enclosures were filled with natural stimuli with those that only had artificially enriched enclosures (e.g. a hollow log and a plastic igloo served the function of providing shelter in the separate cages) (Lambert, Nelson, Jovanovic, & Cerdá, 2015). While there were no differences found in intelligence between the two groups of rats, the researchers observed 'biomarkers of emotional resilience' in the rats exposed to the natural objects (p. 109). The 'country rats', as this group became known, also showed 'increased levels of exploratory behavior' and 'interacted with all of the elements of their environment (i.e. physical stimuli, other rats and the cage itself)' when compared to the 'city rats' (p. 110). In fact, when I saw the images of the two group of rats included in Lambert and colleagues' article, I couldn't help notice that the 'country rats' reminded me of Ellen's classroom environment while the 'city rats' look like most school classrooms I have seen.

Lambert's (2008) earlier research highlights the ways that human emotions developed in the context of our evolution as a species, and should actually be considered evolutionary adaptive responses. She calls attention to how our brains evolved in complex environments, in which our ancestors' survival depended on constant physical interaction with their natural surroundings. That is, humans did not evolve independently, but rather, by interacting *with* other creatures, with the earth and skies, the winds and rocks.

Most importantly, Lambert's (2008) research shows that our bodies, senses and emotions were crucial to human evolution and becoming, but today increasing numbers of people all over the world are engaging almost exclusively in cognitive tasks, while essentially ignoring the body, which she points out makes us sick and depressed.

Jackie:

This conversation makes me think about why I prefer the term 'ecology' to 'environment'. The latter can reinforce the divisions between what is human and not-human. So, Camden is taught to recycle and compost, and made to feel guilt and shame about this, yet you also told me a horrifying story about a time when he was not allowed to go to the bathroom when he needed to at school. His need to urinate is a fundamental biological requirement. Even at the level of the body, 'nature' is denied in favour of the 'mind' and 'discipline' (so he should learn to hold it, right?!). Schools here are currently obsessed with 'brain' science, in their never-ending quest for constant progress and improvement on irrelevant rating scales (increasingly irrelevant given the magnitude of challenges and troubles the human community is facing). But what does his body know and learn through his schooling? So, I prefer the term ecological because the prefix 'eco' signals relationships and interconnection, as well as points to the importance of diversity of all kinds and the danger of monocultures (whether for agriculture or children in schools). Maybe we should preface 'environment' or 'critical' with 'eco'… an eco-critical environmental education! Or a critical eco-environmental education. Or a critical environmental eco-education.

If critical environmental education is ecologically oriented, it has to engage and take into account the body and body experiences in the world. It would necessarily illuminate the relations, for example, between growing food and food security or between healthy and unhealthy soil. I recently read that the United Nations is predicting up to another 700 million refugees by 2050 due to soil erosion and degradation, with all the related conflict and hunger and suffering that comes with that (Watts, 2018). You suggested to Camden's teacher that she grow food in their classroom. Some schools do grow food, but most do not. Done thoughtfully, this would be a significant first step in creating a different kind of education, far beyond guilt and panic-inducing recycling. If children were given the opportunity to understand what it means to nourish the body, to enjoy eating together, to create community around shared circumstances with diverse others, to practice gratitude together for all the life that sustains us, and so on, I feel like maybe we have a chance to protect some future possibilities for life on this planet. Imagine that!

Stephanie:

Another important and related aspect of Kelly Lambert's (2008) work is that she brings attention to the work of neurobiologists, who speculate based on the large amount of brain space dedicated to hand movement that the human brain evolved as our ancestors performed tasks related to survival that required extensive hand use, such as hunting, gathering, gardening, cooking and making clothing. Indeed, if human body parts were proportional to the amount of brain space devoted to their function, the body would take on a completely different shape: a proportional figure would consist of enormous hands, a large mouth, a nose, two eyes and almost no body (see drawing, p. 85). Some scientists speculate, too, that because the brain's language centre and the areas that control the hands are found close together, the development of language 'may be intimately connected to conveying information about the intricate use of the hands to generate and use tools' (pp. 121–122).

Jackie:

Exactly! Schooling children out of the knowledge that a branch is an E seems really risky and dangerous in this time.

The fragmented and separative (Keller, 1986, 1996) ways of thinking give rise also to similar ways of approaching something like species extinction. So maybe, we plant certain plants in our school yard to 'save bees' or 'save butterflies', but a much more difficult concept is to link our own human living to the whole interdependent web of life. There is no saving bees or whales or polar bears or butterflies or all the various species whose names we do not know, without saving ourselves, or vice versa. It's all one ecological whole. Therefore, an (eco)critical (eco)environmental (eco)education would be an education that works to undo ALL the life destroying false divisions and dichotomies that schools continue to foster (and fester). These include grade levels and ability levels, subject disciplines, etc., as well as the unequal power structures between adults and children/young people, and between teachers and upper levels of administration. As long as those remain, any kind of meaningful environmental education seems close to impossible, as it merely becomes one subject discipline competing for dominance amongst many.

I wonder if a way to heal these relations (heal has the same roots as whole, hale, health, etc.) is through our breath, connected to that branch, and through the 'E' taking on the quality of something sacred and precious, not something that is reduced to some micro or phonetic unit.

Stephanie:

This reminds me of the moment in Kate Horsley's (2001) novel, *Confessions of a Pagan Nun*, when the main character shares her realization that all things are sacred 'but in ways that the human cannot understand with thoughts but must know in the *moment between breaths*' (p. 163, emphasis added). I have always taken comfort in the ways that breathing connects humans to one another and to the natural world. When we breathe out air molecules, some of those same molecules are subsequently breathed into the lungs of others, including those of our adversaries, of animals, of birds and even of insects! Plants take in the carbon dioxide from our breath and return oxygen to the atmosphere so that we may continue this breathing—and life itself. Our breath co-mingles with the air, it swirls around and brushes up against 'inanimate' rocks and dirt.

Jackie:

In this sense, the E holds the universe. It is not reducible to rules and procedures—for example, sit still, hold your pencil properly, destroy the world, language goes extinct! These are also relations that are important to be able to 'see' and be critical about. As teachers, caught up in the historical structures of schools, it is very challenging for us to gain insight into these processes and our own entanglement in their violence. In those squeezed and tight structures of schooling, there is scant time or room for branches. In the grown-up world of rushing and work and debt and consuming, a branch is *never* an E. Also, there is constant panic about 'literacy', narrowly defined as reading and writing skills, where children are seen as 'behind' if they don't know their E's (or ABC's) at age 4 or 6 or wherever our current benchmark and developmental obsession lies.

Stephanie:

Exactly. Camden's school recently purchased and introduced RazKids, a for-profit, online reading program developed in the United States. I was horrified as I began to review the poor-quality, borderline racist materials and I requested that he be removed immediately from the program. Upon asking friends across the country, I realized that this program has been purchased and is being used in many public elementary schools across Canada.

Jackie:

This is a profound example of the ways that capitalism is increasingly making its way into public education in a way that undermines any possibility of *critical* environmental education. The panic about 'literacy' opens the door for programmes like RazKids to infiltrate schools, with its confident promises to swiftly progress children through reading 'levels' and propel them into serving the global economy. Even if the programme is patriarchal and racist, and takes away from children their ability to choose for themselves what they would like to read and learn, somehow it becomes justifiable.

Maybe, this is where the 'critical' comes in. Prefacing 'environmental education' with 'critical' means that complex, interdisciplinary understandings would be fostered as a *first thought*. Things are not neatly recycled or recyclable. The demand for cheap, disposable goods has deathly consequences, for some humans more than others, and for some creatures more than others. 'Critical' signals the necessary and difficult work of seeking to understand historical, cultural and economic entanglements in a system that is not designed to support life. I think capitalism (or now, neoliberal economic globalization) is completely thrilled when we spend our time trying to save bees and whales, especially when it involves buying more things in the name of the 'saving', even when what Donna Haraway (2016) calls the *ongoingness* of species, or life itself, is completely threatened. This is not about making rules for recycling and then going on to torture children with a mass-marketed racist levelled reading program from Texas! Indeed, we should understand the connections between these. Both are life destroying. Buying such programs with public tax money is consumerism gone insane. It is based on greed and fear, and it provides profits to a corporation that cares neither for children nor the world.

An (eco)critical environmental education/curriculum/practice would refuse to participate in such programs because they are NOT about the branch being an E (or actual reading or learning for that matter), and in fact, such programs depend on forgetting ALL such relationships. Because if we REMEMBERED them, these programs would lose their relevance and power. We'd see them for the ridiculous and dangerous things they are.

Stephanie:

It was you who introduced me to the work of Catherine Keller (1990), who reflected that 'a grounded self, unlike a fixed ego, thrives in its dependence on earth and only as earthling, on the matrix of relations to all the other earthlings' (p. 221). I think that storytelling has the power to infuse environmental education with exactly this—that we are all earthlings, who live and die like all other beings. As a parent, I wonder: what would it mean if children in schools were learning to see and understand themselves as *earthlings*?

Jackie:

I love that! A curriculum for earthlings. Precisely what's required for this time.

References

Abram, D. (1997). *The spell of the sensuous: Perception and language in a more than human world*. New York, NY: Vintage.

Arendt, H. (1968). *Between past and future: Eight exercises in political thought*. New York: Penguin Books.

Bratman, G. N., Hamilton, J. P., Hahn, K. S., Daily, G. C., & Gross, J. J. (2015). Nature experience reduces rumination and subgenual prefrontal cortex activation. *Proceedings of the National Academy of Sciences of the United States of America, 112*(28), 8567–8572. https://doi.org/10.1073/pnas.1510459112

Crouse, D. L., Pinault, L., Balram, A., Hystad, P., Peters, P. A., Chen, H., et al. (2017). Urban greenness and mortality in Canada's largest cities: A national cohort study. *The Lancet Planetary Health, 1*(7), e289–e297. https://doi.org/10.1016/S2542-5196(17)30118-3

Haraway, D. J. (2016). *Staying with the trouble: Making kin in the Chthulucene*. Durham, NC: Duke University Press.

Hartig, T., Evans, G. W., Jamner, L. D., Davis, D. S., & Gärling, T. (2003). Tracking restoration in natural and urban field settings. *Journal of Environmental Psychology, 23*(2), 109–123. https://doi.org/10.1016/S0272-4944(02)00109-3

Horsley, K. (2001). *Confessions of a pagan nun*. Boston, MA: Shambala Publications.

Jensen, D. (2004). *Walking on water: Reading, writing, and revolution*. White River Junction, VT: Chelsea Green Publishing Company.

Keller, C. (1986). *From a broken web: Separation, sexism, and self*. Boston, MA: Beacon Press.

Keller, C. (1990). 'To illuminate your trace': Self in late modern feminist theology. *Listening (Fall), 23*, 211–224.

Keller, C. (1996). *Apocalypse now and then: A feminist guide to the end of the world*. Boston: Beacon Press.

Keller, C. (2002). Interview with Catherine Keller. In D. Jensen (Ed.), *Listening to the land: Conversations about nature, culture, and eros* (pp. 273–281). San Francisco, CA: Sierra Club Books.

Kühn, S., Düzel, S., Eibich, P., Krekel, C., Wüstemann, H., Kolbe, J., et al. (2017). In search of features that constitute an 'enriched environment' in humans: Associations between geographical properties and brain structure. *Scientific Reports, 7*(11920), 1–8. https://doi.org/10.1038/s41598-017-12046-7

Lambert, K. (2008). *Lifting depression: A neuroscientist's hands-on approach to activating your brain's healing power*. New York, NY: Basic Books.

Lambert, K. G., Nelson, R. J., Jovanovic, T., & Cerdá, M. (2015). Brains in the city: Neurobiological effects of urbanization. *Neuroscience and Biobehavioral Reviews, 58*(Nov), 107–122. https://doi.org/10.1016/j.neubiorev.2015.04.007

Lederach, J. P. (2005). *The moral imagination: The art and soul of peacebuilding*. New York, NY: Oxford University Press.

Neilsen, L. (1998). *Knowing her place: Research literacies and feminist occasions*. San Francisco, CA: Caddo Gap Press.

Nettle, D., & Romaine, S. (2000). *Vanishing voices: The extinction of the world's languages*. Oxford: Oxford University Press.

Nieuwenhuijsen, M. J., Khreis, H., Triguero-Mas, M., Gascon, M., & Dadvan, P. (2017). Fifty shades of green: Pathway to healthy urban living. *Epidemiology, 28*(1), 63–71. https://doi.org/10.1097/EDE.0000000000000549

Norris, J. (2008). Duoethnography. In L. M. Given (Ed.), *The SAGE encyclopedia of qualitative research methods* (pp. 233–236). Los Angeles, CA: SAGE.

Norris, J., Sawyer, R. D., & Lund, D. (Eds.). (2012). *Duoethnography: Dialogic methods for social, health, and educational research*. Walnut Creek, CA: Left Coast Press.

Okri, B. (1998). *A way of being free*. London, UK: Phoenix House.

Sawyer, R. D., & Norris, J. (2013). *Duoethnography*. New York, NY: Oxford University Press.

Seidel, J. (2007). *Unknowing the future: Meditations on the time of teaching*. Unpublished Ph.D. dissertation, University of Alberta, Edmonton, Canada.

Seidel, J., & Hill, L. (2015). Thinking together: A duoethnographic inquiry into the implementation of a field experience curriculum. *In Education*. Retrieved from http://ineducation.ca/ineducation/article/view/219/800

Taylor, A. F., Kuo, F. E., & Sullivan, W. C. (2001). Coping with ADD: The surprising connection to green play settings. *Environment and Behavior, 33*(1), 54–77.

Watts, J. (2018, March 26). Land degradation threatens human wellbeing, major report warns. *The Guardian*. Retrieved from https://www.theguardian.com/environment/2018/mar/26/land-degradation-is-undermining-human-wellbeing-un-report-warns

Chapter 5
Integrating Struggles for Environmental Justice into the Curriculum: A Critical Pedagogy Viewpoint (*three Case Studies from Greece: The Keratea's Sanitary Landfill, the Chalcidice's Gold Mines and the Struggle Regarding Asopos River)

Aristotelis S. Gkiolmas

Introduction

Being a main aspect in the curricula of many European and Northern American countries, environmental education (often referred to as EE) is more or less rigidly situated in these curricula, obeying a vast range of aims and perspectives.

There has been a lot of discussion and a lot of educational literature on what should be—and what should not be—included in the environmental educational curriculum, in the preschool, the primary, the secondary or even the tertiary education (Aikens, McKenzie, & Vaughter, 2016; Figueroa, 2002). But, even if there are very interesting and important suggestions on this, there seems to be a shortage of existing work (Hodson, 2014; Smith & Sobel, 2014) on a specific component: the inclusion of the environmental justice activism and the environmental struggles in the environmental education curricula (Adamson et al., 2002; Washington & Strong, 1997).

As a natural outcome, the question follows on why such an infusion is considered as necessary, since the students' programmes of studies are already overwhelmed with a lot of work on a variety of fields (science, maths, language and history among others).

The first reason for this embodiment is the self-evident tendency for the diffusion of environmental justice's knowledge to the wide public (Pezzullo & Cox, 2017). Struggles for environmental justice stem from—but also produce knowledge about—the environment, and the classrooms are the most appropriate loci to diffuse such knowledge.

A. S. Gkiolmas (✉)
Department of Primary Education, University of Athens, Athens, Greece
e-mail: agkiolm@primedu.uoa.gr

A. S. Gkiolmas, C. D. Skordoulis (eds.), *Towards Critical Environmental Education*, Critical Studies of Education 14,
https://doi.org/10.1007/978-3-030-50609-4_5

The second is that, among the main demands of environmental education, is to motivate the students about specific environmental issues. What should be a more proper way to do so, than to involve them (theoretically) in an ongoing environmental justice struggle in their own land?

As a third reason emerges the link between the school and the society. It would be inconceivable to exclude primary, secondary and university classrooms from such an important matter for the society (local or global) as is an environmental justice struggle unfolding out there.

At first, a distinction is necessary between *environmental movement* and *environmental justice struggles*, as they are meant in this chapter. The environmental movement—for example, ecological movement—is an ongoing effort of the people and intellectuals in many countries, in order to address various environmental issues that create burdens in their lives (Gottlieb, 2005). The environmental justice struggles, on the other hand, are an aspect of this movement, referring to the hard efforts of social groups, tribes, genders or classes to remove the *unfair* distribution of environmental problems over them. Usually, they are 'episodes' with a specific time and space range (Martinez-Alier, Temper, Del Bene, & Scheidel, 2016). It is rather difficult, seemingly, to define something as a *global* environmental justice movement.

A Historical Overview and a Framework of Both Fields (EE and Environmental Justice Struggles)

It has to be noted that although the form of environmental problems has changed a lot in the last decades, such an evolution and transformation is not observed in environmental education, at least in Europe and North America. Old problems, such as the greenhouse effect, become acceleratingly more and more serious. Additionally, new problems are accumulating, such as the garbage in space, the genetically modified organisms (GMOs), the radiation of mobile phone antennas, among others. Apart from that, the accumulation of wealth to the few and the spread of poverty to most people are even more extreme in the last decades, a phenomenon which is at the core of environmental problems.

Usually, environmental struggles do not stem from a rigid theoretical basis, because simply they are not meant to do so. They stem from people, or groups of people, such as people of colour, women or habitants of indigenous areas, who feel that they are threatened to such a point that they need to organise and act accumulatively.

Environmental justice emerged as a term in the United States in the 1980s. The United States' Environmental Protection Agency (EPA) gives a first definition of the term:

'Environmental justice is the fair treatment and meaningful involvement of all people regardless of race, colour, national origin or income with respect to the development, implementation, and enforcement of environmental laws, regulations, and policies. EPA has this goal for all communities and persons across this Nation. It will be achieved when everyone enjoys the same degree of protection from environmental and health hazards and equal access to the decision-making process to have a healthy environment in which to live, learn, and work'.

Many authors consider the riots of 1982 in Warren County, North Carolina, as the starting point for the environmental justice movement in the United States: A poor, and predominantly African-American, county was chosen by the State of North Carolina for the placing of a toxic waste landfill, in order to dispose of polychlorinated biphenyls (PCBs), illegally dumped along the roadway of 14 counties. Community residents enlisted the support of the United Church of Christ Commission for Racial Justice to engage a campaign of non-violent civil disobedience which ultimately resulted in over 500 arrests. Warren County is, therefore, widely viewed as the juncture which marks the transformation of environmental problems in people of colour and low-income communities from isolated issues into the present national concern around environmental justice.

Outside the United States, in the rest of the world, contemporary environmental justice struggles are considered to be strongly connected with the anti-globalisation struggles that made their appearances in conferences such as Seattle 1999 or Porto-Alegre 2004. They mainly address the major issue of the unequal distribution of goods and resources, as well the exploitation of local communities by large multinationals, aided by trademark globalisation institutes, such as the International Monetary Fund (IMF), the WTO (World Trade Organisation) or the World Bank (Scholsberg, 2007).

Another major issue in the environmental justice movements around the world has been the unequal effects of the climate change. Certain non-governmental organisations (NGOs)—not all of them, since the role of some of them is dubious—as well as initiatives in Third World countries, such as the International Climate Justice Network (ICJN), the Environmental Justice and Climate Change Initiative (EJCCI) and the Rising Tide, are pressing continuously, historically, for climate change to be stopped, or—at least at first—to be equally horrific for all people on Earth.

Other initiatives are giving environmental justice struggles for the Native Americans in both North America and South America (the Coordinator of Indigenous Organizations of the Amazon River Basin, COICA, being a typical and among the most successful examples) (Scholsberg, 2007).

In India, Vandana Shiva (2015, 2016), the famous eco-feminist and ecological justice activist, describes, in her works, the various battles having been given there for the preservation of cultural identities against the environmental pressures, as well as for the water, the food supply network inequalities, etc.

The Environmental Justice Atlas (2018), an international collaboration that tracks land and energy conflicts around the world, has spotted the ten most important environmental justice battles being active nowadays all over the world. These struggles include land grabbing (Asia and Central and South America), renewable energy conflicts (Asia, Africa, South America), unburnable fuels (mainly European countries concerning the Arctic and South American countries) and trash economy

(GAIA, the Global Alliance for Incinerator Alternatives, unites communities resisting incinerators, mainly in countries of Asia Pacific and South America).

If someone considers which are *the main ways/tools* by which the reasons for an environmental justice struggle, wherever on Earth, the evolution of it, as well as the outcomes of it, would reach the broader audiences, these might be, up to our days:

(a) The press (written or digital).
(b) The mass media.
(c) The Internet (social media, blogs, etc.)
(d) The direct oral communication.

One thing that should be stressed is that very rarely either (1) the classroom or (2) the teacher's instruction is—until now—included in the above means.

The Necessity for Bringing Environmental Justice Struggles into the Curriculum and the Educational Praxis

As they are grounded nowadays, the European and Northern American curricula bear very little relation to the environmental activism. What is meant by environmental education is mainly a body of knowledge concerning the environment and the ecosystems, and—on the other hand—a collection (or collections) of activities in daily life or in the field that could lead to a better understanding of the environmental problems or to a better treatment of the environment on behalf of the human beings (in this case, the students).

But it is a widespread acceptance in the world—and especially in the politically alert citizens—that environmental knowledge and environmental action have, now, mainly to do with economics and politics.

So, first, what is happening to the environment and to the natural resources worldwide is mainly a result of political decisions and of financial interests and strategies.

It is unthinkable that the students (of the primary and secondary grades) are not informed and are not aware of it.

Second, the environmental issues and burdens are not equally distributed all over the Earth or—in many cases—all over the same country, and this is in the core of environmental justice education. For example, how can it be possible that students in China and students in the United Kingdom learn about the 'greenhouse effect' in the same manner and within the same settings? For the former, this is a problem that alters—and usually destroys—their lives, whereas for the latter in many areas, it does not even become apparent.

Third, when the stakeholders and the policymakers of educational systems do speak about 'active citizenship education,' they usually exclude environmental justice education. There is no agenda for students becoming more aware of environmental justice in the typical educational declarations of the European Union, the US government or the UNESCO brochures.

A final argument, in favour of introducing directly the struggles for environmental justice into the taught syllabi, is that classrooms represent a microcosm of the society and the social context around them. So if any kind of struggles are born and carried on into the society, they should be directly reflected in the classroom praxis.

By itself, education about environmental justice struggles can enrich courses such as (a) a course on history, (b) a course of any kind of environmental science and (c) a course on sociology. Furthermore, this could be reflected in the interdisciplinary activities and the taught material.

The Direct Connection Between Environmental Justice Struggles and Education: Overview of Various Curricula

Many environmental justice struggles are inherently connected to classroom education. Students and teachers participate in one of the struggling parts, usually. The schools, or university campuses, are very often the places where people meet and discuss how the struggle is going up to that point, or what its future evolution should be.

Apart from that, historical outcomes and ethics of such struggles are often discussed by teachers in topics like history, sociology and citizenship.

An important issue to be discussed is that the simple reference of and discussion about environmental justice struggles within the school (or university) classrooms is not considered as adequate.

The need to involve environmental education with environmental justice activism stems from various resources (Scandrett, 2012; Strife, 2010).

It is an apparent consequence that there should be a direct reciprocal link between EE and the justice struggles. More well-educated students of all levels (primary, secondary, etc.) lead to better and better theoretically grounded activists. Reciprocally, the outcomes of the struggles do enrich what is taught and learned about environmental justice in the classrooms.

So in Fig. 5.1, we propose what is a main feedback-loop-scheme in our current work.

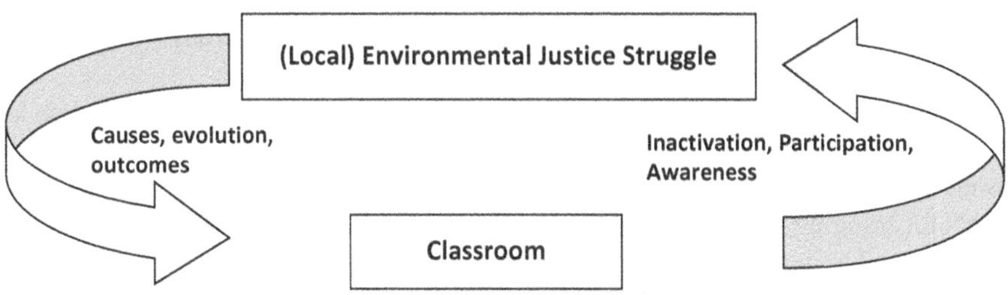

Fig. 5.1 The ongoing loop between an environmental justice struggle and the classroom

In the United Kingdom, the National Association for Environmental Education (NAEE) has published 'The Environmental Curriculum' (NAEE, 2018), which contains general, as well as more specific, guidelines on how to educate students about the environment, both in the classroom and in field training. It also gives instructions about how specific fields (physics, chemistry, biology, etc.) can be related to environmental education. However, it does not contain any reference or prompting to involve environmental justice issues and movements in the things being taught.

In the United States, as a general line, the environmental education curriculum is supervised by the Environmental Protection Agency (www.epa.gov), which—nevertheless—gives a lot of freedom to local educators and schools to follow their own lines of instruction in the field, and just provides resources or model lessons and activities. Another resource from the EPA in United States, the 'Environmental Justice 2020 Action Agenda' (EJ-2020 Action Agenda), states *three goals, eight priority areas, and four national challenges, which are very loose and generally determined and not 'harmful' at all, for those who create the problems. A new strategy towards creating Curricula with more connection to Environmental justice activism are the so called 'co-created' Curricula in UK and the USA* (Bovill, 2014).

As far as Europe is concerned, an overview of the educational policies is followed, as regards the Environmental and Sustainability Education was published (Jucker & Mathar, 2015). There it becomes evident that within the European curricula for Environmental and Sustainability Education, the term 'justice' is practically only meant as 'social justice' and 'ecological justice,' which are different terms than environmental justice.

The Key Role of Critical Pedagogy in the Infusion

Involvement with environmental justice activism exists in the core of critical pedagogy. McLaren (2013) has embodied environmental justice in its discourse about eco-pedagogy. Richard Kahn addresses the issue of teaching environmental justice in his work about handling the various crises in our planet through education (Kahn, 2010). Nocella II, Ducre, and Lupinacci (2017) have pointed out the very deep and strong relationship between the lack of environmental justice in everyday life and in schools with very heavy social problems, such as youth incarceration. Even the notion of 'place-based' environmental education, which appears in critical pedagogical environmental views (Gruenewald, 2003), is by itself strongly interwoven with local environmental justice struggles.

Techniques of teaching and learning that are central in critical pedagogy, such as ethnographic research (Denzin, 2017) and the technique of 'bricolage' (Kincheloe, McLaren, & Steinberg, 2011), can be used—as will be shown in this work—as main tools for bringing environmental justice activism into the school-class settings.

The need to involve environmental education with environmental activism stems from various resources (Strife, 2010). What is proposed in this chapter is that critical pedagogy can play a central role into this infusion. At first, in every aspect of education, thus also in environmental education, critical pedagogy refuses the 'banking approach to knowledge' (Freire, 2018). Under this view, the students do not simply learn about the framework of the environmental justice struggle going on in their land, but they do critically intervene with the facts, striving to attain their own views on the issues being posed.

Furthermore, the role of the teacher (the primary *s*chool *or* the *s*econdary *s*chool teacher) is extremely augmented, in the sense of a 'transformative intellectual' (Giroux, Sandlin, Burdick, & Darder, 2018). The role of the teacher in the classroom is not purely to present the acts and the main arguments of the environmental justice struggle to the students, but to transform the way the *latter* react to the facts and to the ideas. The ultimate scope is that the students take a very *active*—perhaps a ***leading***—role in the struggle, participating in the discussion in their family, *or* even in the activist actions, if they are older in age.

An *empowered* educator, in the terms defined by critical pedagogy (Kincheloe, 2011), would be ideal for changing the curriculum towards a more environmental-justice-oriented one, as regards the content and the main axes.

Together with that, an anti-positivist turn or an anti-FIDUROD perspective is necessary in the diffusion of knowledge in the classroom, where FIDUROD (Kincheloe, 2008) is the typical form of Western knowledge diffused in the school:

F = formal, I = intractable, D = decontextualised, U = universalistic, R = reductionistic, OD = one-dimensional.

Critical Pedagogy Tools for the Integration of Environmental Justice

The 'ethnographic research' and the 'bricolage' are two key tools of the critical pedagogy theory that should play central roles in diffusing the environmental justice discourse in the school curricula. A third decisive parameter is the stance towards what critical pedagogy calls the 'hidden curriculum' (Whitty, 2018).

Ethnographic Research

It should form an integrated part of any kind of infusion of environmental justice activism into the school curricula. It is an inherent characteristic of all the environmental justice struggles in the United States, as well as in the rest of the world, that they are case-specific and place-specific. So, it would be decisive for the infusion of this kind of activism into the taught syllabus that ethnographic research is taught and is performed within the classroom settings.

The 'Bricolage'

A mixture of qualitative educational researches, as defined by Steinberg and Kincheloe (2011), is necessary to define the strategy by which the environmental justice activism would be infused into the curriculum. 'Bricolage', as a qualitative form of research, entails incorporating many different perspectives and a mixture of inquiry methodologies, and in particular, according to Kincheloe's (2001) process, it is necessary to include global subjugated and indigenous knowledge. Large samples of students and educators, in the area or the country involved in the environmental justice actions, should be questioned or asked to participate in the bricolage research in various ways, so as to define what would be the proper way to '*teach about*' and '*enact into*' the learners and the educators into the environmental justice activism processes.

Furthermore, one should not forget that environmental justice poses basically the fundamental questions of who suffers environmental problems in their worst expression; and why. But in the existing curricula in the majority of the 'developed' countries of the world, the *hidden curriculum* (McLaren, 2003) prevails, which: (a) implies that the environmental problems, such as pollution or toxic waste, are *neutral*, being the same and of the same extent for all classes and races on Earth; and (b) sees the solution of the environmental problems either as a change of people attitudes or as a technological issue and totally refuting that the unequal distribution of wealth, of power and of the use of natural resources all over the Earth is a basic cause for the environmental injustice. As a typical example, even hurricanes (such as Katrina) or other extreme weather phenomena would not be of the same strength and tremendous results, if they had hit richer and more powerful areas of the planet. Critical pedagogy totally refutes and opposes the bad aspects of the 'hidden curriculum' stating that the structures of hierarchy of rich over poor and of inequality should be spotted in the former and be opposed and changed, in order to address the issue of environmental justice within educational praxis.

The Proposed Methodology of Incorporating the Struggles into Curriculum and Instruction

The *five important steps* suggested, for the inclusion of environmental struggles in the EE curricula, are as follows:

Detailed Information About the Ongoing Struggle or the Battle(s) Being Completed

Any sort of curriculum concerning environmental justice activism should give detailed descriptions of what has been done in the field (both historically and at the contemporary period), in the country or in the area being concerned. There is absolute need of a fair description of the events and an outline of which were the causes that led to the need for environmental justice battles in each case.

Investigation of the Learners' Attitudes about Both Sides Involved in the Struggle

Specific questionnaires are, or can be, designed, in order to check what exactly the learners do know about the environmental struggle(s) taking place in their own homeland. Separate activities could be designed in order to make them come in touch and analyse the struggles taking place in their area. Obviously, the latter also includes field activity.

Presentation of What Each Final Outcome of the Battle Would Mean or Means

The students definitely need to be presented with an—as much as possible accurate and fair—description of what *each one of* the—usually—*two* outcomes of an environmental justice battle would mean for their land and their lives in it. In other words, learn what would be the impact to their own lives, in the case of the prevalence (or "victory") of either the one or the other of the two opponents in the struggle.

Of course, the students have the capability of learning the views about the outcome from the points of view of the two conflicting sides, and this has to do: (1) with the oral descriptions coming from the local society, (2) with the Press and how it presents and analyses the facts, (3) with the Internet and even (4) with the propaganda mechanisms used by one or from both conflicting sides.

But when the presentation of the final outcomes comes from within the classroom and is made by the teacher himself/herself, thus when it is embodied within the official body of the school culture and praxis, this is a totally different thing. When the presentation of the outcomes comes from within the school system, and it is balanced, then the students conceive it as official, authentic knowledge, rigorous and well documented and they are much more willing to hear and to learn or to reproduce in their surroundings, what the two final outcomes of the ongoing environmental struggle would mean for them and their own ways of life.

As a further argument, the action of presenting the impact of the struggle's outcome to the (primary and secondary) students is an action of environmental justice by itself! If, for example, a local tribe of indigenous people in North or South America are striving to remove a very serious toxic waste problem from their homeland, their voices cannot reach the ears of the pupils with the same ease that the government's arguments will reach. So the teacher does justice to them by presenting the *two* possible outcomes to the students, and this fair and balanced presentation is actually environmental justice.

Analysing and Summarising the Things to Be Learned by the Specific Struggle

It should not be forgotten that this chapter refers mainly to primary and secondary education students, and not to the tertiary education students, as have done many researchers and publications in the past (Figueroa, 2002; Holifield, Chakraborty, & Walker, 2017; Washington & Strong, 1997).

Thus, it is a central aspect of this process to summarise *what* has been learned from the scrutiny of the ongoing environmental struggle in the students' own land.

Distributed in the form (1) of worksheets or (2) as a project involving interviews with the citizens being involved and recording of the answers/discussions, the summarisation will aim at focusing the students on results that may be used as guidelines in their future lives, as environmentally active citizens.

The things that can be learned by students, from a specific ongoing environmental struggle, are a rich variety of things, ranging from pieces of knowledge to things concerning politics, their attitude towards nature and the political stances they would cultivate and that they would develop as citizens.

Each environmental struggle is *always* place-specific and time-specific, so the things that the students can learn by being taught about it strongly depend on the area that they live (and into which the struggle unfolds), as well as to the period of time that this happens. But the things to be learned are really important things anyway. They concern the environmental and cultural heritage of the students, the natural resources as the latter are handled by the students' parents, and as these resources will be handled in the future by the students themselves. Also the things to be learned address the political and economic interests that exist in pupils' land in their specific time-spot, interests that drive one—or both—of the conflicting sides.

On the other hand, environmental battles share some universal and not time-restricted characteristics, which can help the students learn crucial things by being taught about the former: (a) There is always *one* side that is environmentally heavily burdened compared to the other. (b) There are always political and environmental conflicts—serious ones—lying behind the environmental conflict. (c) There is always one side that will be politically and economically benefited by the heavy environmental discrimination. (d) The criteria of the discrimination—the injustice—are more or less the same in all cases (colour of skin, gender, nationality, belonging to a special ethnic or social group which is marginalised for certain reasons, class and economical stratification). (e) Notions and concepts such as sustainability, prosperity, human health and use of natural resources are valid for the whole of the human kind, and they cannot appertain to certain people while not appertaining to others.

Reaching Certain Conclusions

It is a central step of the methodology presented here, point to conclusions. The conclusions refer both to the teacher/instructor and his success or failure to infect the classroom curriculum with elements of environmental justice, as well as to the educational stakeholders, as regards their will and ability to do the same.

To point to conclusions, detailed research has to be done to the students but to the teachers as well, regarding the formers' ability to incorporate elements of the environmental justice activism into their body of situated knowledge and daily life. At the same time, research—educational research—is necessary to be performed concerning the extent to which the teacher succeeded in fulfilling the objectives posed, that is, how much did they manage to bring issues of environmental justice struggles into the teaching praxis, and whether they themselves are adequately acquainted with these notions.

The conclusions that both the instructors (teachers) and the students will reach, concerning the environmental justice struggles, are not always quite apparent. A crucial factor is *if there is actually an ongoing struggle (or a struggle that is about to start or it terminated recently)* in the classroom's own area or land. In some cases, *there may be no* highly projected area in the specific locus, so the students are taught about another struggle nearby or far away. In this case, the conclusions that they will reach will be different than those that they would reach if they lived within this struggle. So a different strategy of scrutinising the conclusions would be needed.

Another important factor to examine the conclusions reached by the students when they have been taught about an environmental justice struggle is the *time-distance* from it. If one performs educational research in order to see what has remained in the learners' minds after being taught about the struggle, this should better be performed in two time stages: one shortly after the instruction (and possibly the environmental struggle itself) and one at a later time, possibly 1 or 2 years after the instruction. It is expected that the results about the conclusions reached by the learners will differ significantly. When the struggle has terminated (if it is not a very long-lasting one), the conclusions reached, within the instruction about it, are strongly related to the student's personal (or sentimental) involvement in it. But when there is a significant time-distance from the battle, the conclusions reached by the classroom audience or the students are much more rigid, more clear and more beneficial (in the learning field). Thus, the latter conclusions (time-distant from the instruction and the environmental justice struggle itself) are of a much greater importance, I believe, for the educational researcher.

As already said, there are a variety of ways to extract what the conclusions about the environmental justice battle struggle are, by using tools of the field of educational research. Even addressing the local society of the *adults and* not necessarily the students themselves, an experienced researcher can deduce what the children have learned about the environmental justice struggle.

All these kinds of conclusions reached within the instruction process within the classroom are a very important guiding tool for re-designing the curriculum and the teaching objectives, always having in mind that the central aim is to infuse environmental justice struggles into the classroom material. Of course, an inhibitive factor here will always be that the educational stakeholders, and thus the persons that play a key role in forming the curricula, will—in many cases—belong to the one (the economically strong) side of the struggle, but this is where the role of the national and local movements of people comes. It is everyone's duty to act and play a role in the educational curricula for their own children and for the children of all his/her country. This can be done through the channel of public discourse, or if the governments close such a channel, this can be done through direct activism.

The Future Directions and the Expected Outcomes After Dispersing Environmental Justice into the Curricula

Bringing environmental justice struggles into the primary and secondary education curricula is expected to act as an 'agent' for engaging students into environmental justice activism. The students will become part of the situation evolving around them (their territory, their village or town, their country). As a result, this gives a new boost to the environmental justice struggle taking place in the specific territory and makes them part of the situation.

Additionally, the curricula take a brand new—more activist-based—form compared to what they are now, which is more or less a collection of knowledge and activities about the environment.

Needless to say that this has the re-writing of the school (primary and secondary) curricula in Western countries as a pre-requisite, where the basic principles of environmental justice and environmental justice activism are included. Some such principles are underlined here:

(a) All forms of life on Earth are equal as regards their rights, and if—in some cases—these rights are conflicting, it is *not* the work of humans to resolve the issue.
(b) There should be equal environmental burden, if necessary to all people on Earth, regardless of their income.
(c) No persons should face human-induced environmental disasters in the future, regardless of their colour, race, nationality, gender and religion.

Greece as a Case Study

In Greece, a lot of environmental justice struggles are taking place (Velegrakis & Frezouli, 2016), but three struggles have been chosen as having an interest from a worldwide perspective, as well as having the ability to be enclosed into educational curricular settings.

The first is the struggle for the creation or not of a sanitary landfill in Keratea, Attica (Accardo, 2011). The second is the establishment of gold mines in Cassandra, Chalcidice, on behalf of the Canadian/multi-national company 'Eldorado Gold' (Velegrakis & Frezouli, 2016; Alevizopoulou, 2013), and the third is the poisoning of Asopos River Basin, in Thebes, Central Greece, by the industrial companies which send their waste there (Tentes & Damigos, 2012).

Each of these three cases of environmental justice struggle is different, but all three share common characteristics: They are very intense struggles, with enormous pressure and even violent action from both sides involved. They all raise issues of environmental justice because the extensive environmental burden—sometimes threatening the public health or even the people's lives—falls onto Greek people of the agricultural class, living in the country and not in large cities, possessing usually low income or, as in the case of Keratea, belonging to the poor suburban bourgeois class. All three struggles also have lasted for many years and are still ongoing, not with equal intensity, with Asopos's being the longest in time. Let us have a look on each one separately:

(a) In Keratea, the Greek government decided to build a sanitary landfill, in order to dispose waste from the region of Attica, around—and including—Athens. The choice of the specific place was never adequately explained. The citizens—mainly belonging to the lower bourgeois class and already living in a degraded environment—revolted, and the riots starts in 2008. A picture of the daily clashe[s is depicted in Fig. 5.2. There were fights with the police squads almost

Fig. 5.2 Riots in Keratea (Velegrakis & Frezouli, 2016)

daily and 24-h occupations of the broader area around the place where the landfill was supposed to be built. Now, the riots seem to have abated, but the situation remains unclear, since the government neither started building the landfill there, nor did they announce a new place for its construction.

(b) In Kassandra, Chalcidice, the Greek government agreed with the Canadian/multi-national company 'Eldorado Gold' that the latter would create there a mine for the extraction of gold. They would pay large amounts of money to the government and also—they claim—they would create many work positions for the local population (Alevizopoulou, 2013). On the other hand, such a mine would totally destroy the landscape and the natural ecosystems of this area of Chalcidice peninsula, also threatening the people's lives and making it impossible for them to live around there. Again, belonging to the agricultural class, most of the peasants revolted—a small part of them was in favour of the mines and there were everyday clashes with the police and demonstrations. In some cases, the police used smoke bombs and tear gases, even in areas around schools or against old people. A typical situation of those days is depicted in Fig. 5.3.

Again in this case, the situation remains dull, since the mines have not begun to be built, but there is no official withdrawal of the company from the area.

(c) The situation in Asopos River is different and much older. An industrial area was unofficially established in the territory around the river in the decade of 1970s. The industries gathered there dropped all their toxic waste within the river basin (especially chromium) gradually polluting it tremendously.

Fig. 5.3 A policeman in a Kassandra's village (Velegrakis & Frezouli, 2016)

The percentage of cancers and other diseases into the habitants' population increased dramatically (Leontopoulos, 2013). The people living in the banks of the river started protesting more and more. Even the famous environmental activist Erin Brockovich got involved. After years of protests and legal battles, the situation has reached today to a point where the whole area is characterised and delineated as 'official industrial zone', and thus, the companies are obliged to obey the legislation about industrial pollution. It is dubious, though, if the companies do so, and who is checking it. In Fig. 5.4, people of the area of Asopos are shown, during a demonstration.

The Elements That the Greek Educational Curriculum May Include from the Aforementioned Struggles

The Greek curriculum or the secondary—or in certain aspects the primary – education could be enriched with many aspects and outcomes of the three aforementioned environmental justice struggles.

Firstly, science syllabus could adopt the serious issues and objectives of learning the scientific issues of waste disposal, the issues of rivers' chemical pollution or the chemistry of the burdens that gold extraction causes.

Secondly, the choice of specific peoples' groups and of certain areas when a big work destroying the environment is designed could be included in areas such as sociology, economics and 'study of the environment' (the last being taught in

Fig. 5.4 Protest (one of the numerous) of inhabitants of areas near Asopos River (Rizospastis, 2009)

the Greek primary school). The causes that lead financial stakeholders and environmental government agents to address these people can be studied and taught.

And finally, the struggles for environmental justice as collective and active expressions of communities (mainly rural but also in cities) could form an example of collective movements or—at least – actions to be emphasised and given as examples in the Greek curriculum.

A Methodology of Four Stages for Including These Struggles in the Classroom's Praxis

Being integrated into the Greek educational curriculum, the three aforementioned struggles, as examples of environmental justice and activism, they can as a further step, become a part of the daily educational praxis within the Greek classrooms.

Here, a methodology and a framework for doing this are suggested, including *four stages*, always in close relevance with the principles of—applied—critical pedagogy.

First Stage

The students of the secondary (but with the necessary adoptions and simplifications, even of the primary) schools are given specific questionnaires and worksheets, concerning the struggle. This refers primarily to the schools of the areas neighbouring to the battle, but also all over Greece. The questions are as such:

1. *Do you know what is the situation with* (e.g.) *the landfill in Keratea?*

   ____.

2. *Which are the two—or the three—parts involved in the conflict?*

   ____.

3. *Which is/are the reason(s) for the conflict?*

____.

And so on, with similar questions following, which engage the students with the issue and shed light on their knowledge of it.

Second Stage

The students are prompted, with the intervention of the empowered teacher, to perform ethnographic research but also to use 'bricolage' methodologies, in order to:

(a) Record the views and the actions of the local minorities who are 'suppressed' by the problem.

(b) Record the views of the opposing side (the industries' owners, the government officials, the police, etc.).
(c) Try to see what are the interests 'behind' the struggle, for all sides involved.

In the second stage, the students use 'field action' tools such as the interviews, the written questionnaires, the narratives of local people, the newspapers and the websites.

Third Stage

The students would try to reach conclusions on what the situation is up to that point in time, who is wrong and who is right (if there is one), how different the situation is being presented in the media compared to what the reality is and what the possible final outcome seems to be. This should lead to a discussion in the session of the classroom.

Fourth Stage

A theatrical play or a dramatized episode can be created, with the students performing as actors, some of them playing the revolting people, others playing the police, others the government officials, others the industrialists/investors, etc. It is important that the whole play's or dramatized game's dialogues are written by the students themselves.

Having performed all these four stages, the students are—supposedly—ready to get out in the real field of the environmental justice struggle and participate in it, with the means that *they* can use and to the extent feasible for them.

The teacher, as a transformative intellectual, might also participate, mobilising at the same time the whole group of his colleagues and, possibly, parents who remain inactive until then.

With Greece as a case study of applying all the above, the infusion of environmental justice struggle in the curriculum and in the daily school praxis is getting—hopefully—a guide on how it could be done and how it could flourish properly, with critical pedagogy as a central shaping factor.

References

Accardo N. (2011). The Attican landfill war: A load of trash? Retrieved November 20th, 2018, from http://www.cafebabel.co.uk/culture/article/the-attican-landfill-war-a-load-of-trash.html

Adamson, J., Evans, M., & Stein, R. (2002). *The environmental justice reader: Politics, poetics and pedagogy*. Tucson, AZ: University of Arizona Press.

Aikens, K., McKenzie, M., & Vaughter, P. (2016). Environmental and sustainability education policy research: A systematic review of methodological and thematic trends. *Environmental Education Research, 22*, 333–359.

Alevizopoulou, M. (2013). *Gold mining in Halkidiki: Greek governments in the service of mining companies*. [Greek language]. Adapted from a report published in *UNFOLLOW Magazine*, issue 14. Retrieved from http://borderlinereports.net/2013/02/23/report-gold-mining-in-chalkidiki-part-1-greek-governments-in-the-service-of-mining-companies/

Bovill, C. (2014). An investigation of co-created curricula within higher education in the UK, Ireland and the USA. *Innovations in Education and Teaching International, 51*(1), 15–25.

Denzin, N. K. (2017). Critical qualitative inquiry. *Qualitative Inquiry, 23*, 8–16.

Environmental Justice Atlas. (2018). *The world's top 10 battles for environmental justice*. Retrieved December 10, 2018, from https://cosmosmagazine.com/geoscience/the-world-s-top-10-battles-for-environmental-justice

Figueroa, R. (2002). Teaching for transformation: Lessons from environ-mental justice. In J. Adamson, M. Evans, & R. Stein (Eds.), *The environmental justice reader: Politics, poetics and pedagogy*. Tucson, AZ: University of Arizona Press.

Freire, P. (2018). *Pedagogy of the oppressed* (50th anniversary edition, with an introduction by Donaldo Macedo and an afterword by Ira Shor). New York, NY: Bloomsbury Academic

Giroux, H. A., Sandlin, J. A., Burdick, J., & Darder, A. (2018). *The new Henry Giroux reader: The role of the public intellectual in a time of tyranny*. Gorham, ME: Myers Education Press.

Gottlieb, R. (2005). *Forcing the spring: The transformation of the American environmental movement (revised edn)*. Washington, DC: Island Press.

Gruenewald, D. A. (2003). The best of both worlds: A critical pedagogy of place. *Educational Researcher, 32*(4) , 3–12.

Hodson, D. (2014). Becoming part of the solution: Learning about activism, learning through activism, learning from activism. In L. Bencze & S. Alsop (Eds.), *Activist science and technology education* (pp. 67–98). Dordrecht: Springer.

Holifield, R. B., Chakraborty, J., & Walker, G. P. (2017). *The Routledge handbook of environmental justice*. London: Routledge.

Jucker, R., & Mathar, R. (Eds.). (2015). *Schooling for sustainable development in Europe*. Cham: Springer.

Kahn, R. (2010). *Critical pedagogy, ecoliteracy and planetary crisis: The eco-pedagogy movement*. New York, NY: Peter Lang.

Kincheloe, J. L. (2001). Describing the bricolage: Conceptualizing a new rigour in qualitative research. *Qualitative Inquiry, 7*(6), 679–692.

Kincheloe, J. L. (2008). *Knowledge and critical pedagogy: An introduction*. Rotterdam: Springer B.V..

Kincheloe, J. L. (2011). In K. Hayes, S. Steinberg, & K. Tobin (Eds.), *Key works in critical pedagogy*. Rotterdam: Sense Publishers.

Kincheloe, J. L., McLaren, P., & Steinberg, S. R. (2011). Critical pedagogy, and qualitative research: Moving to the bricolage. In N. K. Denzin & Y. S. Lincoln (Eds.), *The Sage handbook of qualitative research* (pp. 163–178). Los Angeles, CA: SAGE.

Leontopoulos, N. (2013) *Asopos river basin: Heavy metal pollution may be a white collar crime*. Retrieved October 12, 2018, from http://www.thepressproject.gr/details_en.php?aid=51591

Martinez-Alier, J., Temper, L., Del Bene, D., & Scheidel, A. (2016). Is there a global environmental justice movement? *The Journal of Peasant Studies, 43*(3), 731–755.

McLaren, P. (2003). Critical pedagogy: A look at the major concepts. In A. Darder, M. Baltodano, & R. D. Torres (Eds.), *The critical pedagogy reader*. New York, NY: Routledge Farmer.

McLaren, P. (2013). Seeds of resistance: Towards a revolutionary critical ecopedagogy. *The Social Studies, 9*(1), 84–108.

NAEE (National Association for Environmental Education) (2018). *The environmental curriculum*. [Author Julette Green]. Retrieved October 20, 2018, from http://naee.org.uk/wp-content/uploads/2018/02/NAEE-Secondary-Environmental-Curriculum_FINAL.pdf

Nocella II, A. J., Ducre, K. A., & Lupinacci, J. L. (2017). *Addressing environmental and food justice toward dismantling the school-to-prison pipeline*. New York, NY: Palgrave Macmillan.

Pezzullo, P. C., & Cox, R. (2017). *Environmental communication and the public sphere* (5th ed.). Thousand Oaks, CA: Sage Publications.
Rizospastis ("Radical"). (2009, April 24). *Asopos River: The tragic consequences from its pollution*. [Official newspaper of the Greek Communist Party. Online version]. Retrieved December 3, 2018 https://www.rizospastis.gr/story.do?id=5051842
Scandrett, E. (2012). Social learning in environmental justice struggles. In B. Hall, D. Clover, J. Crowther, & E. Scandrett (Eds.), *Learning and education for a better world* (pp. 41–56). Rotterdam: Sense Publishers.
Scholsberg, D. (2007). *Defining environmental justice: Theories, movements and nature*. Oxford: Oxford University Press.
Shiva, V. (2015). *Earth democracy: Justice, sustainability, and peace*. Berkeley, CA: North Atlantic Books.
Shiva, V. (2016). *Who really feeds the world? The failures of agribusiness and the promise of agroecology*. London: Zed Books.
Smith, G. A., & Sobel, D. (2014). *Place- and community-based education in schools*. New York, NY: Routledge. (e-book by Taylor & Francis).
Strife, S. (2010). Reflecting on environmental education: Where is our place in the green movement? *The Journal of Environmental Education, 41*(3), 179–191.
Tentes, G., & Damigos, D. (2012). The lost value of groundwater: The case of Asopos river basin in Central Greece. *Water Resources Management, 26*(1), 147–164.
Velegrakis, G., & Frezouli, E. (2016). *Environmental conflicts and social movements: 12 Greek cases*. Athens: Department of Geography, Harokopio University. [Greek language] Retrieved November 11, 2018, from https://www.researchgate.net/profile/Giorgos_Velegrakis/publication/309385698_Environmental_Conflicts_and_Social_Movements_-_Twelve_case_studies_from_Greece/links/580d045508ae2cb3a5e3c2a2/Environmental-Conflicts-and-Social-Movements-Twelve-case-studies-from-Greece.pdf
Washington, R., & Strong, D. (1997). A model for teaching environmental justice in a planning curriculum. *Journal of Planning Education and Research, 16*(3), 280–290.
Whitty, G. (2018). *Sociology and school knowledge: Curriculum theory, research and politics*. Oxford: Routledge.

Chapter 6
Education for Environmental Equity and Justice: A Graduate Degree in Urban Environmental Education

Cynthia Thomashow

In designing the Urban Environmental Education program, I unraveled and reknitted a long history of academic work in environmental education. For years on end, Antioch University graduated mostly white "nature-based" educators into the EE field. They were well intentioned and resolved to make a difference in the world and most lacked a multicultural and multiracial perspective of environmental issues. A few of my own powerful experiences in which I worked with urban youth and teachers convinced me that most of the approaches and intentions of traditional environmental education were simply not relevant to their lives. An essential perspective of the urban social network was missing.

My environmental work, our work, had to expand into the places where most people live. The mostly "nature-based" environmental field would benefit from a deeper and broader understanding of the intersectionality of environmental issues impacting our cities. Representation from the highly diverse racial and cultural demographic of our cities was missing from our workplaces, our audiences, and our efforts to engage change-makers. Power lies in our multicultural solidarity. The environmental field is challenged to figure out how to become more widely inclusive of all people, places, and realities.

"Nothing About us Without Us is For Us" (African proverb) adorns the t-shirt given to each student upon graduation from Antioch University Seattle's Urban Environmental Education M.A.Ed. program. The 15-month academic experience reflects this sentiment throughout. We pride ourselves on pioneering pedagogy shaped by an ethic of listening from the ground up before settling on solutions to the complex social and ecological issues. The UEE program operates in a cohort model. All of the students take all of the classes together. At the end of the first quarter, a

C. Thomashow (✉)
Antioch University Seattle, Seattle, WA, USA
e-mail: cthomashow@antioch.edu

A. S. Gkiolmas, C. D. Skordoulis (eds.), *Towards Critical Environmental Education*, Critical Studies of Education 14,
https://doi.org/10.1007/978-3-030-50609-4_6

safe space is created for sharing difficult issues of race, equity, and inclusion within the environmental and educational fields. As one of the UEE partners expressed:

> *The UEE program provides a brave space to really lean in and make a difference by holding the promise of bringing front and center the stories of historically marginalized and disenfranchised people and their ways of knowing the earth. Daylighting those stories and voices provides the consideration that they deserve and that we need. The work of UEE expands the brain trust of ways of knowing the earth and respecting multiple voices.* (Belinda Chin, Seattle Parks and Recreation)

Our students learn how to engage community members in conversations that unearth their "real" concerns. Each student is embedded in a community-based organization for 30 weeks, which provides critical knowledge about how people live, work, and play in urban places. The students learn to encounter head-on matters of health, safety, housing, and access to food that impact environmental integrity and well-being. The relationship of social justice and environmental leadership informs all aspects of this groundbreaking program. The UEE program reworks traditional "nature-based" approaches. We thread urban ecosystem dynamics into each quarter of study uncovering how ecological integrity is shaped by politics, economics, diversity, and equity.

Each closely knit student cohort is as diverse as the urban communities we study and serve. This rich diversity ensures authentic and visceral representations of how race and inequities impact experience with pollution, gentrification, access to healthy food, waste management, lack of green space, and the impacts of climate change. For many of our students, it is their first experience in a classroom with a majority people of color. For most of our students, UEE is the first academic experience with mostly instructors of color. For our white students, it is the first "environmental" set of courses where they may be in the minority.

We are breaking the "green ceiling." What does that mean? Racism and inequality, displacement and exclusion are as important to the environmental conversation as urban ecology and educational strategies. The faculty work hard to create a safe space for reflection and direct confrontation of these facts without sacrificing attention to issues like climate change, biodiversity loss, or waste streams. The students work with each other - in and outside the classroom - to find truth, clarity, and a way forward that integrates the social/cultural and environmental realities. Environmental issues are laced with social exploitation, unfair health and wellness policies, disparities in pollution, the lack of inclusion in urban planning, and more. Once we work through initial raw and angry feelings, our goal is to find a way to work together with respect and understanding.

This chapter will provide examples of some of the educational approaches designed by the faculty in the UEE program. The faculty are challenged to prepare future environmental leaders to grapple with the environmental issues in urban habitats where most of the human population lives. Our approaches and strategies are innovative and responsive to a changing climate and dense, complex urban centers. We are out on the streets and in communities as much as we are in the classroom.

The chapter will also capture some of the learning reflections penned by UEE students. We encourage our students to think critically, to develop cultural competency, and to cultivate educational approaches that embrace equity, inclusion, and collective action. UEE faculty work together to examine the deep intersections of ecology and infrastructure, politics and place, social justice and environmental integrity as the basis for research projects, learning, and praxis.

As we work to extend the focus of environmental education into city streets and step into the places where most of us live, justice issues become an integral part of understanding the dynamic forces that shape urban centers. Why do some have greater access to health, shelter, work, food, and green spaces than others? How does the movement and settlement of people influence the ecology of a place? How does infrastructure and new development influence the quality of water, air, access to food and shelter, displacement, and homelessness? Instituting "ways of knowing" that intentionally frame "environment" through the urban experience considers the reciprocal relationship between the social and the ecological.

We start with what's around us: the people close to home and the issues they are facing. We work with schools, communities, local organizations and systems to understand the most important factors that come with living in an urban environment and what can be done to create positive change. CJ Goulding, 2016 graduate, employed by Children and Nature Network Natural Leaders

As educators, long-term results rely on building Trust among constituents, learners, and community members. First, we build relationships...authentic and real relationships. Relationships are key to the longevity of any environmental solutions. We need to step outside of our personal assumptions, our biases, our stereotypes and listen to the stories from inside a community. The real experiences of everyday people shine a light on the environmental issues they face. Embedded in those stories are the keys to building stewards of urban places. (Jess Wallach, 2016 graduate, employed by 350.org)

The Future of Environmental Learning

The foundational UEE course, The Future of Environmental Learning, takes a critical look at the field and its necessary transformation. Mitchell Thomashow's course introduces new ways of thinking about environmental issues, widening the traditional scope of the discipline. In his own words, *"Feverish tides of change are sweeping the globe. The same tensions that converged in the 1960s (Civil Rights, Feminism, Peace, and Environment) inflame educators as a result of the very environmental issues we've been worried about—scarcities prompted by natural disasters, political upheavals, and religious extremism; refugees forced to abandon their homes and livelihoods; interconnected global economic cycles that result in demographic dislocations and migrations; with the attendant impacts on ecosystems and the biosphere."*

One of the first activities in this class introduces "sense of place" and the formation of ecological identity. Understanding one's ecological identity is a critical aspect of environmental work and intention. Sense of place activities get at the heart of how one's identity fuels the nature and quality of environmental leadership. The description from his syllabus reads:

Sense of place is a search for ecological roots. This is best accomplished when you have a relationship to the land on which you live, when you can place yourself securely in a tangible place. It's through the place that you live that you construct your personal identity, your relationship to the landscape, and you determine what is important in your life. Sense of place concerns your home and region, feelings about land and community, kindred species, community niches, and sacred places. To have a sense of place is to merge your personal geography with the ecological landscape, incorporate maps of memory with how you dwell in a bioregion.

The sense of place map is a rite of passage that links ecological identity to life cycle development. What are the feelings, events, and choices that characterize how you see yourself in the biosphere through different periods of your life, through various dwellings and travels in time and space. How will you communicate and illustrate the places where you've been, where you live now, and where you see yourself in the future? (Ecological Identity, Mitchell Thomashow, MIT Press)

Powerful student exhibitions provide the venue to share multiple dimensions of one's connection to the biosphere, to each other, to land and history, to culture, race, and politics. The "maps" serve as provocative expressions of value and intention that expand the sense of what connects all of us to ecosystems, environmental issues, and the tumultuous issues of race, culture, and justice.

Thomashow invites the students to consider how environmental leadership must evolve to include the "tides of change":

For the field of environmental studies, including approaches to activism and education, it's not sufficient to work mainly in the area of conservation and environmental protection. To remain pertinent and responsive, environmental citizens must demonstrate how these convergent challenges are inextricably linked to the fate of the planet. Let's reframe the tides of change as questions:

1. *The rapacious exploitation of the biosphere and its life systems continues unabated. How do we best communicate the necessity of ecosystem thinking?*
2. *There is an increasing disparity between rich and poor. How do we promote economic equity and social justice in cultures of materialism and entitlement?*
3. *There is great apprehension concerning the integration and separation of global cultures. How do we promote intercultural understanding and cosmopolitan thinking in the midst of nationalist responses and ethnic tribalism?*
4. *Violence, weaponry, and terror compete with deliberation, diplomacy, and collaboration. How do we settle our differences through community democracy, service and compromise, in the midst of conflict, extreme behavior, and fear?*

The emerging environmental movement of the early twenty-first century has a new shape and form. Visit any grassroots community-based environmental project and speak to the staff and participants. You will find a young generation of activists who understand the necessity of working in diverse communities to promote constructive change. They are concerned about environmental issues, but they are equally committed to approaches that emphasize diversity, equity, and inclusion. They are flexible users of social media and they utilize these skills to build coalitions and promote their ideas. (Mitchell Thomashow, To Know the World: A Vision for the Future of Environmental Learning, MIT Press, 2020)

Migration is raised as a seminal feature of environmental change. Following the sharing of "sense of place maps," each student creates and shares a chart of their own family's history of migration. One student may talk about a farming heritage that has rooted their family to a particular place for generations while another may talk about spending a childhood moving up the coast of California from Mexico picking fruit; yet another may have fled political terrorism and come to the U.S. for asylum. Intentional relocation, unresolvable displacement, purposeful migration… all of the students have an experience of moving, uprooting, leaving family, and establishing a new sense of place. This activity reveals how common the experience of migration is to humankind; how humans and animals are often on the move to escape drought, ecosystem changes, and climate impacts. The study of migration is linked to the dynamics of climate change, of power dynamics, of health concerns, and political upheaval, and all are related to environmental integrity.

Urbanizing Environmental Education

Students are anxious to take their new awareness into the streets. "Urbanizing Environmental Education" converts what they have learned theoretically into action. They all investigate the mechanics of the city, how power and money determine the shape and access to private and public spaces, how and why movement through the streets is managed, the ways that water, waste, and wildlife are moved through places, where the displaced have gone, "red-lining" impacts on community make-up, where food can be accessed and why some have green space and others do not. All of these factors become educational priorities. The students are encouraged to invent strategies that organize the engagement of communities in learning the answers to questions that impact their lives.

I teach this class. There are "big ideas" hiding in the streets of cities that are central to environmental learning. Identifying and articulating these important concepts is critical to shaping effective civic engagement and environmental education. Big Ideas lay the groundwork for educational activities. Here are a few from this summer:

- The design of parks and green spaces impact the health of communities.
- Public artwork serves a political purpose.
- Cultural integrity is maintained through architecture and design.
- *Structural urban development influences the movement, settlement, and displacement of people.*
- Private property influences the access to public spaces.

Activities are developed to unpack these "big ideas." We go out into the city to try them out. As an example, one group of students settled on creating activities that focus on the concept "private property influences access to public spaces." We walked down seven blocks of Seattle's Belltown neighborhood with notebooks in hand. As historical background, we learned that as recently as 10 years ago, Belltown was a city-sponsored free bus ride away for anyone in need. That has changed. Now, Uber and Lyft are your choices for transport.

The class researches the settlement, gentrification, and displacement of inhabitants within the Belltown area through narratives and photos. Belltown in the 1990s was filled with low-income housing, services for unemployed and homeless, a destination for mental health services, and shelters. Some of the people who used these services still call this home but now find shelter in the alleyways. In 1994, I hired three "houseless" men from the "Millionaire Club," a charity that finds part-time employment for men who are homeless, to guide a different group of students through the Belltown section of the city and interpret their experiences in their own words. This year none are available.

Our instructional walk is guided by questions: Observe what you observe. That may sound trite. However, most of us do not see a lot when walking city streets. We encourage students to change their focus…how would a small child navigate this street? What would someone in a wheelchair or on crutches be looking at? How would you describe the Belltown neighborhood from the perspective of wildlife? What do you think has changed over the past 10 years? What evidence of past "eras" remains? How has the increase in private residential property influenced the shape of the streets? And more…. at the end of our exploration, reflective essays reveal things that have never been considered before. Here is an example of one reflection:

> *On Vine Street in the Belltown neighborhood, a community group living in the new upscale high-rise apartment complex has transformed the city block into an urban watershed oasis. The intention is admirable; to slow the flow of storm water into the Puget Sound. The new resident group has been influential in shrinking two lanes of traffic on the north side of the building into one in order to accommodate a new bio-swale gardens. This is an environmentalist's dream! The "Cistern Steps" is a series of terraced plantings designed to clean rainwater as it travels through the city. It is an oasis for several reasons. One is aesthetic. Amidst the heavily built environment of Belltown, these green terraced plantings echo an English countryside garden and has changed the pattern of traffic to a trickle. Street noise is diminished. Greenery is abundant and leads into a protected community garden full of food and flowers. I was immediately impressed by the beauty and function of the Cistern Steps, until I noticed on very odd feature of the design. The Cistern Steps are* ***steps****…meaning that people in wheelchairs, parents pushing strollers and the disabled, cannot use this city block. It got me wondering about the line between public and private. One block uphill from the "steps" is another stretch of plantings that used to be a wide sidewalk. A "No Trespassing" sign sits planted amongst ferns, shrubs, and storm-water pools surrounded by a spiky metal fence to discourage sitting, loitering, sleeping. A "No Trespassing" sign on a public city street should make us all start thinking: what kind of influence, power, and position allows for this line to exist between public and private in a city, who gets the right to control accessibility to a public walkway? (Melani Baker, UEE Alum, 2019)*

Of course, the discussion following this activity ends up very passionate as the relationship between social justice and environmental reveals itself. One observation indicates that as the "environment" becomes greener and cleaner, "unsavory" people are moved out. This lesson provides an example of how those with power, privilege, and money can initiate "environmental improvements" that speed up the attraction of this area to developers. What of the displaced? Was there any effort put into soliciting input from the original tenants? Might their ideas have created a way to coexist in this place? How might an educator engage community in thinking through an inclusive lens to improve a neighborhood? Discussions following "lessons" like this one lead to students unpacking the swirling undercurrents of racism, class issues, power, money, and privilege in urban planning. Once the relationships between environmental change and social justice are unveiled from this concrete experience on the street, the faculty can build the bridges to the role educators might play as community advocates, change agents, interpreters, and facilitators.

Race, Equity, Inclusion, and Environment

The focus on race, equity, and inclusion in relationship to environmental issues is what makes our program different from others. The diversity of the student body creates a microcosm mirroring the challenge of engaging people in hard conversations. Students spend 30 weeks in three courses that study the relationship between race, equity, and environmental issues. Multicultural education and strategic leadership follow the foundation class. Faculty work to peel back layers of history, life experiences, and identity in order to facilitate learning that builds skills in community engagement and collective impact. The following reflective essay captures an awareness gained in class that changed a career. The transformative awareness from this activity was volcanic. Josh became, after graduation, an educational leader facilitating Race, Equity, and Inclusion initiatives for public schools. This new awareness shaped his UEE research, which exposed the difficulty of making public lands accessible to people of color when the interpretive narratives exclude multicultural perspectives.

> *One strong memory that I carry is trying out an activity that had us fill out a racial privilege survey. 12 students out of 18 in the UEE cohort were people of color. The activity positioned us around the room according to our answers. I'm sure that it was designed to enlighten the white people present. It set a tone for acknowledging that we are not all experiencing the "environment" in the same way, that we have* ***not*** *all been treated equally, that equity is a huge issue in the environmental field (define environment any way you like here). There are historical impacts on each of our relationships with "environment." We are each living a legacy that shapes our understanding of how things are, who we are in them and who has the power to act. It caused me to pause. Here I was, beginning to bond with a new group, as a "white" man in a diverse group of fellow colleagues. I had to acknowledge that I didn't experience "environment" in the same way that everyone else did. My assumptions crumbled. I had been wrong in my assumptions as an educator previously. I began to realize that what had been missing from conversations about environmental issues was the impact of oppression and white supremacy in the environmental field. It's one thing to have faculty of color and another to work with peers who are coming from a huge range of identities, experiences and backgrounds that differ significantly from your own. We began sharing personal and professional experiences, identities, emotions and vulnerabilities in the safe place that was provided by the program. Hearing honest and clear representations of each other's journey and how racial identity influenced each person's work in the field was profound. And it was bounded by classes where theorists and writers like Bell Hooks and Paulo Freire held us up and provided the words to underwrite and translate this difficult topic into a reality to use in our work. (Josh Parker, UEE alum, 2018)*

The Practicum

The practicum is at the heart of our academic program. Early on, Sue Byers, the current director, and I decided that the traditional short dips into community organizations (that most graduate programs design) would not work for us. Our students spend 30 weeks with one community-based organization in a paid contracted position. "Nothing About Us Without Us Is For Us" guides research, the creation of educational events, the organization of civic engagement, and the facilitation collective action. Paolo Friere and Bell Hooks are in our back pockets:

> *Education is a mechanism for social change... it becomes the "practice of freedom", the means by which men and women deal critically with reality and discover how to participate in the transformation of their world. (Gramsci, Freire, and Adult Education: Possibilities for Transformative Action, by Peter Mayo, Macmillan, 1999)*

The UEE program prides itself on the quality and number of practicum partnerships that have been created and nurtured over the last 5 years. Our students engage in collaborative research and community engagement projects within communities all over the city…governmental, national, service-oriented, health advocacy, food access, housing, etc. They are actively engaged in organizational life while wearing the "hat" of a participatory action researcher, all the while supported by weekly classes and supervisory oversight from UEE faculty. Our students are managing projects within The National Park Service, Seattle Parks and Recreation, The National Forest Service, Brightwater Treatment Plant for King County, Seattle Public Schools, Beacon Hill Food Forest, El Centro de la Raza (The Center for People of All Races), White Center Development Association, and so many more. Issues raised in classes are heightened and practiced in practicum settings. Sylvia, a woman of color, wrote this reflection after a meeting with Seattle Parks and Recreation supported her learning:

> *A few weeks ago, I went out to the beautiful Cedar River Watershed Outdoor Education Center with my practicum supervisor. My practicum title is Green Jobs Research Assistant at Seattle Parks and Recreation (SPR) and SPR environmental learning unit. We were a team of naturalists and environmental educators given the task of planning 2018's approach to environmental education. The day began with some discussion on where diversity is lacking and what needed to be done to improve. It quickly became apparent, like it usually does, that diversity was professional vernacular for race. The environmental educators noticed that most of their students were middle or upper middle-class white students. They wanted to bring more youth of color, immigrants, youth from other economic backgrounds into the environmental learning centers.*
>
> *This was a timely opportunity for me to have a chance to apply what we have been discussing in our UEE classes. I began to think about a book we had been reading in Participatory Action Research class: Paulo Freire's The Pedagogy of the Oppressed. In Freire's theory, the leaders of a group and the students must work together to learn from each other in loving humility. There cannot be a sense of us vs. them. There is only the we, who work together toward a common goal, in this case, environmental justice and equity. To me, it seems like the goal of inclusion of "othered" populations, the environmental educators would first need to start by "un-othering" them in their minds, that is, deconstructing the us vs. them mentality. Othering, by definition, is the act of a dominant group mentally classifying another group as "not one of us."*

Back to the meeting. As I was about to comment on these reflections, one of the other UEE grads took the thoughts right out of my mind. "How are we involving the community to understand what they need and how we can best serve them?" she asked. I qualified her statement, commenting that "at times in meetings like this I find that I am a part of the "them" that is being talked about rather than the "us" who may be doing the planning." The other UEE alum then posited, "I wonder if that is how the people we want to serve feel...?"

The room was silent for a few moments. Heads began nodding. What I learned was that there already was a sense that the community needed to be involved in environmental program planning, but that day there was a greater awareness to the radical idea of "we," instead of "us and them." I left the meeting very excited to see the theory that I learned in the classroom put into practice in a real-world situation with real implications in the city that I call home. (Sylvia Hadnot, UEE Alum, 2018)

The relationship between social justice, city life, and environmental conditions is complex and it all becomes very real in the practicum placements. Here are some of the recent Legacy project titles:

- Cultivating equitable youth leadership within a wilderness society
- Increasing access to outdoor adventure programming to youth of color
- Bringing communities of color and neighbors into neighborhood green-spaces on their own terms
- Including counter narratives and indigenous perspectives in National Forest interpretation
- Digital storytelling as a method to advance the focus of youth on environmental issues
- Implementing inclusive community input into the development of NPS interpretive signage
- Increasing racial and gender equity in STEM-based girls programming

The Urban Environmental Education Master's program at Antioch University Seattle intentionally embeds our students within urban communities to learn first-hand about the threads that hold a place together and the forces that make living conditions and environmental realities what they are. Each practicum leads to a Legacy project that stems from research and practice. Legacy projects provide the information necessary for organizations to move forward on changes and improvements in educational and programmatic design.

Our students find employment after graduation because they have studied, researched, and practiced the integration of social justice and environmental leadership. Eighty-five percent of our graduates are actively working in the field. They rely on each other for support. The cohorts continue as learning communities long after graduation. Here is a final reflection from a graduate, seeking solace from her UEE cohort:

Sometimes, the skyscape in Seattle uncannily reflects our inner landscape. Election day, November 8th, 2016, was gray and cloudy. Politics had brought its own darkness to the future for many of us. Suddenly, the sun seemed so much farther from reach. The shift in the political climate rocked the emotions of our UEE cohort, a group of women and men of many colors, homelands and passions, who all care deeply about the health of people and our planet. All of whom were afraid of the racism and environmental abuse that might flow from this election. Some of us were shocked, some of us not as much. All of us concerned about the future of the environmental landscape, our professional field in light of the transition of power.

We entered an academic map at the cutting edge of Urban Environmental Education. We are drawing the edges of this discipline as we push them. We seek to be a new kind of pioneer, a kind that draws on old and honored knowledge which has ridden the undercurrents for centuries, and on fresh new knowledge sparked from the minds of a truly diverse cohort of learners. We strive to set the table with welcoming places for all communities to speak their mind. We envision a society built on systems that support and regenerate life where all humans flourish because our ecosystems do. And here we are. Front and center in a time when the progress to halt climate change is threatened and the humanity of millions of people is being disregarded.

As a cohort, we are more determined than ever to continue our mission. We come to this work for many reasons. All of us believe that Urban communities deserve access to culturally relevant, educational experiences that connect them to their ecological identity and sense of place. We seek to empower groups of people who have traditionally been left out of the mainstream environmental movement, yet who have often been the most impacted by the reasons the field exists. We know that knowledge is power, that cities are diverse and complicated and that that makes them as interesting and important as any rainforest, savannah or coral reef. We come together with the communities and leaders with whom we practice our ideals. Our study and work confirms our commitment supporting the rights of people and their access to healthy vibrant environments. This solidarity is the light we need to keep going. (Annalise Ritter, UEE Alum, 2017)

Chapter 7
Our Human-Centeredness Is Killing Us: A Case for Indigenizing Diversity Education

Four Arrows, Henry Fowler, Katrina S. Rogers, and David Blake Willis

Insisting that we are sentient, and all others are dumb, is clinging to an arbitrarily narrow perspective that allows us to do great violence to others and, ultimately, to ourselves as well.

(White & Guyette, 2010, p. 113).

Part One: The Roots of Anthropocentrism

Critical Environmental Education is on the threshold of a powerful change with a spotlight now focused on *human education.*[1] Rapidly moving to center stage in the consciousness of serious planners, whether they be civilian, governmental, or military, environmental threats are recognized as calling for an understanding about the environment that reaches beyond simple cause-and-effect relationships. The consequences of not planning and not acting are simply unfathomable, as nearly every indicator of climate change that was predicted appears to have been deeply underestimated. The wisdom and fragility of these connections was long ago appreciated in Indigenous cultures around the planet as diverse as the Kogi of South America

[1] Although education texts such as this one typically use APA style for citations, here in our chapter, we join a symbolic "boycott" against APA for its historical and problematic ethics and anti-Indianism as well as its more recent collusion and cover-up of the illegal torture policies of the U.S. government as exposed in 2015. We are thus opting to use MLA instead. In so doing, we continue to recognize the efforts of APA members to mitigate such problems from the inside and to respect the very positive work done by most APA affiliates and projects. We remain concerned, however, that such actions have not resulted in any significant accountability or sanctions. With great appreciation for this text's editors in allowing us to do this, we recognize that it does not necessarily indicate agreement by everyone in the volume. An important principle of Indigenous Wisdom is to bring many voices around the Fire.

F. Arrows (✉) · H. Fowler · K. S. Rogers · D. B. Willis
Fielding Graduate University, Santa Barbara, CA, USA
e-mail: djacobs@fielding.edu

A. S. Gkiolmas, C. D. Skordoulis (eds.), *Towards Critical Environmental Education*, Critical Studies of Education 14,
https://doi.org/10.1007/978-3-030-50609-4_7

(Ereira, 2012), the Maasai and Kikuyu of East Africa (Maathai, 2003), the Aboriginal peoples of Australia, and the Inuit of the polar regions, among others.

In this chapter, we ask that our readers consider the foundational concepts which have brought us to this stage. We begin with a recognition that both the colonial mercantile and industrial ages have included savage, unforgiveable depredations in the name of God, King, or Queen. In fact, our human greed for treasure, the allure of gold and all its Faustian consequences, has had catastrophic consequences for an array of diverse and lively world societies. Many of these human cultures, somewhere in the neighborhood of 7000–8000, are as endangered as the animals, plants, insects, and other members of our larger Biome, that massive, alive wonder that is our planet.

Our primary focus for the rest of the twenty-first century will be on the problem of *anthropocentrism*, with the goal of this chapter being to emphasize how anthropocentrism is a foundational cause for disrespecting biodiversity and, ironically, diversity among human beings. We submit that our prejudice begins by how we have ignored, dismissed, or side-stepped Indigenous worldview perspectives in education, even in progressive educational movements such as critical pedagogy and holistic education where boundaried relationships between human and other-than-human entities are taken for granted and humans are understood to be separate from nature. Before looking at an alternative thinking and what to do with it, it seems important that we offer a brief history of why anthropocentrism is such a dominant assumption in education today.

The Dilemmas of History: "Civilization" as Problematic, the Anthropocene Revealed

The Anthropocene marks a human-dominated era when our numbers and activities are having dramatic, far-reaching impacts throughout the planet. Like earlier geologic eras described by scientists, the Anthropocene is a particular geologic time, in this case either after or part of the Holocene, which began about 8000 BCE at the end of the last glacial period. What is different is the human-influenced, or anthropogenic, nature of this era. The word combines the root *anthropo* (human) with *-cene*, the suffix for epoch (please see the three sites mentioned in the References that all begin with 'Welcome to the Anthropocene' 2011, 2012, 2013). As *The Economist* has said, "Humans have changed the way the world works. Now they have to change the way they think about it, too."

Here, we highlight the important nexus of environmental education and diversity education, notably the absence of critical reflection in the latter on the differences between Indigenous and dominant worldviews. What are the philosophical, spiritual (worldview), education, and policy orientations that reflect the influence of anthropocentrism? What colonialisms, external and internal, have become deeply embedded in our souls, to the extent that we are literally killing ourselves just as we so profligately overconsume and overextract the bountiful resources of our planet?

Human belief systems before the advent of cities were cyclically based, derived from the rhythms of the seasons, the migrations of animals, and the repeating patterns of the sun and the moon. Orderliness in this prehistory meant a confidence in the return of phenomena and an appreciation of the roles of all players. Life was seen as an ensemble cast and drama rather than one of heroic protagonists and the story arc of an epic journey. Please note the emphasis on his-story, the idea of a heroic journey that begins with the telling of stories, the great epics that are in the beginning cyclical but that are gradually replaced by singular male heroic stories. In this sense, Aphrodite or Isis is gradually displaced by Apollo or Osiris and matriarchally based societies fade or are pushed to the margins by the patriarchal order. Gods come to dominate the theologies rather than Goddesses, and a concern with the existential questions of fate and judgment that parallels the development of cities and civilization evolves into a hierarchy with dominance, a valued and sanctioned approach to society rather than a communal and balanced view.

What we have lost is also an understanding of the role of language, which has become codified and seemingly made permanent by the text rather than the telling and a privileging of the written version of events that empowered the tellers who had access to the code. For most of human history, oral rather than textual traditions were the primary mode of communicating the order of the universe. We can see this with Sanskrit and how it is still handed down primarily as a memorized and repeated oral tradition. Humans begin first with oral traditions and then text, but our era has upended this ordering and practically eliminated the power of the voice.

As humans settled in urban communities, the city (*civ*) became identified with stability, power, and culture, while those people still living nomadically in cyclical patterns and speaking incomprehensibly were beyond the Pale. Without the 'Barbarians,' we cannot have 'Civilization' ('people of the city' as contrasted with the wandering souls or tribes who speak that *ba-ba* tongue), however. Moreover, when cultural traditions became identified with the inerrant traditions of *The Book* (*The Bible*) or the *Analects*, and then later the Charter of the Company (beginning with the Dutch East Indies Company and then the British East India Company), the received wisdom was male, linear, and predatory.

The Age of Colonialism or Imperialism is the beginning of what might also be called the Capitalocene (Fassbinder in this volume and another term for our era; we also highlight Henderson's chapter here on class). Predatory Capitalism utilizes a clear dictum in the Bible, which has been affirmed by not only Christians but Jews and Muslims as well, which clearly indicates and dictates human separation from, and dominance over, nature. *Genesis 1:28*: *God blessed them and said to them, 'Be fruitful and increase in number; fill the earth and subdue it. Rule over the fish of the sea and the birds of the air and over every living creature that moves on the ground.*

Rebirth, the Problem of Sentience, and the Dance of Time

Contrast this with beliefs in reincarnation from India, Amerindians, and elsewhere with the cyclical epistemology so prevalent in the 'prehistory' of humanity. The widespread philosophical support for a human-centered world from ancient to contemporary philosophers is now giving way to a return to pretextual, prehistorical visions. The interplay of the human and plant worlds is spoken of, for example, by the ubiquitous depictions in South Asian art of the beautiful Ashoka tree intertwined with female deities, the *yakshi* or *shalabhanjika*. Frequently seen in South Asian art, belief has it that the tree blossoms only when it is kicked hard by a *shalabhanjika*, young maiden (Tudge, 2006, also remarked in his treatise on trees on this fertility; see also Gupta, 1971, p. 96).

The associations of trees with good fortune and fertility ('touch wood, knock on wood') are of course well known and deeply embedded in Shinto, Buddhism, and Hinduism. It is also related to health and well-being as we see in the Japanese concept of *shinrin yoku* (forest bathing). Sufi and Zen traditions have a similar feel, as revealed in the writings, for example, of Peter Matthiessen (2008) in *The Snow Leopard*, just as Jonathan Balcombe (2011, 2017) has taken us on a similar journey of altering our consciousness about creatures seen as simply created for our exploitation. As with many Indigenous traditions, animals are in fact sentient in the Great Chain of Being, a chain no longer seen as hierarchical and created solely for our exploitation (see Lovejoy, 1933/2010) but as horizontal with interwoven connections, suggesting an ensemble cast instead of a prime hero at the center of the narrative. Seemingly inanimate "objects" like trees and fish take on a whole new realm of consciousness in this understanding (Balcombe, 2017; Wohlleben, 2016a, 2016b).

We note, too, that there is a body of theoretical literature in environmental education about the dangers of anthropocentrism, yet little of this has found its way into "diversity and inclusion" publications, as mentioned earlier. In fact, it is almost entirely absent. Diversity is consistently conceptualized as being exclusively about honoring differences in race, ethnicity, sexual orientation, gender, socioeconomic status, age, ability, unique experiential background, and religious or political beliefs of fellow humans—in other words, human diversity alone. This is despite the fact that many of the great activists of our time like Wangari Maathaai, Krishnammal, and Jagannathan (Coppo, 2004) and other recipients of the Right Livelihood Award (2018, the alternative Nobel Peace Prize) have blended views on the environment and on human societies. Such exclusive human-centeredness in diversity education is not only a barrier to the kinds of environmental education projects John Lupinacci calls for in his chapter in this book, but also it prevents many people from an interest in environmental education in the first place.

It is worthwhile to take a moment to reflect on the significance of religious and spiritual beliefs to understand more clearly the role they play in the maintenance of anthropocentrism. Rebirth or reincarnation has featured in most if not all Indigenous traditions, whereas all the major sects of Abrahamic religions offer inconsistent or unclear ideas about whether or not animals or plants have souls. Even where such a belief exists, humans are generally regarded as superior in the scheme of things. Most have fully dismissed the idea of reincarnation. (Interestingly, the separation of humans and animals was reinforced by the removal of the concept of reincarnation from Christianity at the Council of Nicea in 325.)

Unlike Abrahamic religious beliefs, Buddhism, Hinduism, and other South Asian traditions see humans as having lived before as other-than-human entities and all life forms as sentient, similar to Indigenous Peoples (See *Jatakas* and *Panchatantra*). This "karmic eschatology" in Indic traditions represents a much stronger affiliation with nonhuman life than the oft-stated compassion put forth in Abrahamic traditions. Indic traditional beliefs, of course, also see the human incarnation as *karmic progress* from insects, plants, and animal to human (or in reverse, depending on one's karma). Furthermore, both Eastern and Western religious teachings tend to see escape from the Earth and from earthly existence as the primary goal of life on Earth.

Most Indigenous cultures have rebirthing eschatological beliefs that influence how other-than-human sentient beings are treated that are somewhat different from these Indic ones. Considerations of good or bad karma are rarely a part of traditional beliefs, reincarnations to animal, fish, bird, or plant representing a significant percentage of incarnations, with such reincarnations on an equal footing with human life.

> The constant blurring of human–animal categories and the empathy that humans are expected to have toward animals are based on the important eschatological assumption of "an originally undifferentiated universe in which the boundaries between the human and nonhuman, the spiritual and material, were shifting and permeable," quite unlike the Christian assumption that animals have no souls (Obeyesekere, 2002, p. 45).

In Antonia Mills and Richard Clobodin's scholarly text on reincarnation beliefs among North American Indians, Mills writes that such transmigration would be expected because of "the basic premise that animal life forms are as sentient and evolved as human" (Mills, 2015, p. 34). In fact, many Indigenous cultures see nonhuman sentient beings as teachers, with intimate, protective, and benevolent qualities. Mills believes that rebirthing ideas offered by some contemporary Indigenous individuals see returning as an animal as a punishment "represent[s] a syncretism of Christian belief in after-life punishment and aboriginal ideas of transmutation" (Ibid). Traditional beliefs are, if anything, quite the opposite. For example, Olson describes a Kwakiutl man who remembered having been a salmon in a previous life. This man while fishing recalled having been caught at the same place during his previous life. He even recalled being "cooked, canned, and shipped far away" (Olson, 1949, p. 102).

Aside from eschatological assumptions, Western philosophy itself has supported our current anti-nature and human-centered behavior. Although European thinkers see the Greeks as their intellectual forebears, the pre-Socratic Greeks themselves looked toward the East for the sources of wisdom. In *The Phaedrus*, for example, Plato has Socrates saying he cannot learn out among the trees but only in towns among men (civilization again), yet even Plato believed in reincarnation. Most Greek scholarship responsible for the founding of Western philosophy, however, has focused and continues to focus on ideas that separate human life from nonhuman life. Certain religions, however, especially Jainism, view animals and even plants as endowed with consciousness and belonging with humans to a larger order of sentient existence.

Sustainability and Phronesis

Traditional Indigenous wisdom in many cultures thus focuses on a circular logic of renewal and rebirth, which our recent concern with sustainability has revived. Sustainability returns the logic to our past, more matriarchal understandings of life, and is a powerful metaphor for our predicament as a species in the Age of the Anthropocene. The extent of the challenges and the rapid timing with which Sustainability is teaching us as a metaphor indicate that the time has come to begin the hard work of balancing human needs with those of other species and of our physical and geological home.

We may want to give new attention to the Greek idea of *phronesis*, or practical wisdom, as a way to reveal actionable values, practices, and processes associated with sustainability (Stillman, 2015; see also the Center for Wisdom at the University of Chicago). Values and norms for activism and conservation (LeGrand, 2015) revive traditional understandings of the universe when it comes to environmental sustainability. We face a stark, existential question: *Can societies encourage sufficient changes in human lifestyles to avert ecological collapse?*

What is needed now is a "deep sustainability leadership" of leaders (Schein, 2015) and a motivation to influence behavior and capacity leading to large-scale transformational change. Schein explored this through interviews with 65 Corporate Sustainability Leaders in more than 40 multinational corporations, NGOs, and consultancies. Utilizing fields as diverse as integral ecology, deep ecology, and eco-psychology, he saw the need to move away from "the ecocentric–anthropocentric continuum" to what he calls "the eco-psychological foundations" for thought and action. The need for leadership to move to this vision of ecological sustainability, with the understanding that financial, social, and natural systems are interconnected, has also been foregrounded by Alice MacGillivray (2015). Here, *boundary* is seen as *the* central concept in systems thinking, with an emphasis on how leaders understand and work with boundaries.

The edges of organizations and groups can be seen as places for the mixing of diverse ideas to enable learning and innovation, just as the edges in nature can be places for the mixing of diverse nutrients and species to enable high productivity. On a local level, the emergence of women who are sustainability entrepreneurs as portrayed by Jo-Anne Clarke (2015) gives us pause and hope for enhancing community and improving the health of our planet.

Sustainability is thus about not only the environment, ecological fragility, and balance (viz., Willis, Steier, & Stillman, 2015). Sustainability and this return to Indigenous wisdom threaten the established patriarchal order, not least in the corridors of economic power and transformation. Cultivating a systemic understanding as a holistic way of seeing and acting (so often identified with Indigenous perspectives as we will see in what follows) stands in stark contrast to more industrial-era, reductionistic ways of knowing (see also Skordoulis and Gkiolmas in this book).

This way of understanding appears as cyclical in most cultures regardless of the spiritual or religious orientations. The wisdom of the *Tandava*, the classic Dance of

Shiva that Ananda Coomaraswamy (1918) has described so well, speaks of the power of the moment, of *ichigo-ichie* (once-in-a-lifetime meetings, moments, and chances) as the being and becoming that continually unfolds. Likewise, in China, Japan, and Korea, the back and forth flow of powerful centers alternates with times of fragmented states, continually returning, helping us understand that it is indeed the circle, the cycle, that we are a part of and will always return to as members of human societies. This does not mean that we should accept our fate, a vision that nineteenth-century colonials had of Indians and other Asians ('fatalism') that did a disservice to the agency and action that we do have and take part in. We thus embrace circulation as cultural *process* and search here for new imaginations and, even more critical, interrogations, probing for new vistas and perspectives to help us understand what has been happening (not least in aesthetics, as described in the chapters in this book by Reynolds and Porfilio). This is a reimagining with the twenty-first century in mind, a new era that has moved us to approaches to social phenomena that are multilateral, multipolar, and multicultural. Mixing, mashing, hybridities, and creolizations assert themselves like the destructive yet creative Dance of Shiva, after which newness springs up. But sometimes that which is "new" is actually "old."

Part Two: The Indigenous Perspective

> *Every time you pull a plant from the earth she feels that pull and you must always make the proper offering (so she understands) that what she is giving you is one of the parts of her body* (Cajete, 1994, p. 102).
>
> *The grim prognosis for life on this planet is the consequence of a few centuries of forgetting what traditional Indigenous societies knew and the surviving ones still recognize* (Chomsky, 2013, p. i).

Traditional Indigenous approaches to learning about life skills and values had as paramount the idea of being intimately related to nonhuman life forms. Honoring and learning from animals, plants, birds, bodies of water, and the fish that dwell in them were inseparable from any learning experience and from any life application of the learning. Moreover, in the Indigenous perspective, humans are the younger brothers and sisters of the nonhuman elders of creation and the nonhuman elders are our teachers. In describing how Indigenous cultures see us as "lesser beings in the democracy of species," Robin Wall Kimmerer (2013) asks what the world would be like if educators and scientists saw plants and animals as teachers rather than subjects (p. 347). She, like Noam Chomsky and a growing number of scholars, believes that if we continue to reject this view, our prospects for survival are marginal. This section refers to four Indigenous precepts that are related to living in ways that are not anthropocentric. These include respect for the significance of all life; believing in complementarity of opposites; avoidance of hierarchy; and taking only what is needed from our nonhuman relations.

When one lives in a way where they believe a tree is a relative rather than a resource, diversity and inclusion as related to fellow human-beings follow. Couple this with the concept of Earth and our Mother and with matriarchal cultures, and we can understand why most precontact cultures were "gender-egalitarian and consensus-based societies, actively promoting peace and sustainability by various well-conceived guidelines relating to their economic, social, political and cultural features" (Goettner-Abendroth, 2018, p. 4). The scholars who contributed to Goettner-Abendroth's (2009) earlier edited volume, *Societies of Peace*, reveal that such societies were also nonviolent and practiced great respect for all living creatures without exploitation of humans, animals, or nature. They "revered women as sources of knowledge, cultural progenitors and influential decision-makers because of their ability to bear children, teach the family and contribute to their communities" (Smith, 2008, p. 575.) Prior to European contact, "violence against women was nearly unheard of but was dealt with seriously when it did occur" (ibid.).

Re-embracing such notions of diversity thus requires re-learning a new level of respect for other-than-human life that requires a new level of respect for all life. This requires living according to an Indigenous worldview that psychological anthropologist Hillary S. Webb (2012) refers to as "complementary duality." This contrasts with Western models that see opposites as incompatible "and are therefore engaged in an eternal antagonism and struggle for dominance" (p. 2). This idea of complementarity is foundational to Indigenous Worldview and its understanding that all forms of nature have varying degrees and kinds of sentience. "The Twinned Cosmos pervades every facet of Indigenous philosophy… Everything that exists exists by halves…inseparable and mutually reinforcing…When the Twinship stops, the Cosmos crumbles" (Mann, 2016, pp. 242–243.) Consider the natural symbiotic dynamics of blood and breath, air and water, or male and female.

This is not a claim that everything is automatically in balance. To the contrary, balance points are always shifting. Creative tensions are always happening between complementary opposites. Consider, for example, the tension between the dominant and the Indigenous worldviews as related to human-centeredness. We are not trying to find a balance between them per se but rather to understand where complementarity exists. It is like seeing the differences in blood and breath. Something could go wrong with your blood that is messing with the flowing balance between blood and breath but healing is not really "balancing the two" but finding the problem in the blood that can put it back into harmony with the breath. Imagine the Cherokee Indians in nineteenth century Georgia who adopted settler lifestyles in ways that allowed them to hold onto their traditional beliefs. This allowed for a relatively flowing balance between the cultures until the Euro-Americans discovered gold and then the mindset of materialism, greed, hierarchal thinking went so far in the unhealthy direction and the Trail of Tears resulted. Similarly, we are killing ourselves with the anthropocentric part of our dominant worldview, but viewing the two positions as a strictly antagonistic duality cannot work. We must learn where it goes wrong and do everything to correct the misdirection.

Besides the phenomenon of antagonistic opposites as opposed to complementary duality being a barrier to authentic respect for diversity, the idea of hierarchy prevalent in the dominant worldview is also problematic. Hierarchical worldview leads to domination of others, not inclusion. Harmony does not result from people following the directions of "superiors" but rather from following "Pathways." Cajete describes Indigenous Pathways as a structural metaphor about learning to live life optimally and for the greater good: "In traveling a Pathway we make stops, encounter and overcome obstacles, recognize and interpret signs, seek answers and follow the tracks of entities that have something to teach us" (p. 55). Internalizing what is learned comes from critical thinking, planning, implementation of resultant choices, and reflection for ensuring that the right path is being followed. It does not come from authoritarian mandates.

Furthermore, this learning journey is based on assumptions about the beautiful gift of diverse and inclusive life on Mother Earth. For example, the Navajo believe that good life resides in every angle of the morning light with a promising sense of beauty, hope, and determination for every individual. The Navajo understand that with a sense of the complementary and the supplementary an individual will feel beauty above, below, around, and before, and behind—from every angle. This perspective stems from the natural surroundings—plants, animals, rivers, mountains, sky, etc. These are what bring forth the energy of spirit that aligns with the purpose of life, i.e., to keep in balance, harmony, and respect with the natural order. The only true authority for living life well comes from this spiritual energy, although wise elders who still remember the old ways can help with their knowledge of songs, prayers, ceremonies, and oral storytelling of the Navajo Creation Story, which frames the Navajo epistemology. Ultimately, however, the Navajo epistemology comes from the womb of Mother Earth, Father Sky, sacred mountains, darkness and day, white and yellow corn, Corn Pollen Boy, Beetle Girl, Changing Woman, First Man and Woman, fire, water, and air. It comes from realizing that every aspect of life is an integral part of the whole that includes tiny insects, rivers, trees, mountains, canyons, etc.—all of which are part of the elements that complete the natural order of the universe.

The fourth precept that is vitally related to diversity and inclusion education in ways that are seldom considered has to do with connecting it to an absence of greed. Indigenous understandings include knowing that life must consume life in order to continue, but taking more than one needs and without great respect and expressed gratitude has dire consequences. One of the consequences is developing hierarchy, forgetting complementarity, disrespecting diversity, and restricting inclusion. This is why the taking of animals, birds, fish, and plants for food is accompanied by stories about the wisdom of each life-giver and ceremonies to remind of oneness with them and to give deep gratitude for the sacrifice made. For example, the Pueblo Indian stories teach that each animal has a spirit village. When hunted and killed, their spirits return to it and tell about how they were treated by the hunter. If treated inappropriately, the village might decide to stop giving themselves to humans. The Hopi with a deep understanding of the corn plant and its spiritual power practiced elaborate, artful ceremonies in honor of each stage of planting and harvesting. Whether killing a deer, harvesting corn, picking berries, or digging up roots, we are in effect consuming our relations. Traditional Indigenous cultures prioritized thinking about ways of consumption that are just. Kimmerer, for instance, expresses the idea beautifully when talking about gathering wild berries:

> I live vicariously through the through the photosynthesis of others. I am not the vibrant leaves on the forest floor-I am the woman with the basket and how I fill it is a question that matters. If we are fully awake, a moral question arises as we extinguish the other lives around us on behalf of our own. Whether we are digging wild leeks or going to the mall, how do we consume in a way that does justice to the lives we take? (p. 177).

Such Indigenous perspectives as we have presented are how we lived for 99% of human history (Four Arrows, 2016). They remain in our DNA. Behavioral epigenetics explains how we might have taken a wrong turn long ago but it also tells us that we can reverse direction within a single lifetime. What intentional actions can we do now to start this process that can help us to live in ways that truly see how Indigenous understandings of biodiversity create social systems that reflect great respect for human diversity and inclusion? We suggest a deep and wide commitment to authentic critical environmental education that involves worldview reflection.

Part Three: Critical Environmental Education and Worldview Reflection

> *There is a need to move towards...theory and practice that embraces diverse understandings of knowledge and that recognizes, respects, and builds on pre-existing knowledge systems. This will not only result in better processes and outcomes for Indigenous communities, it will also provide rich learning for mainstream... scholarship and practice.*
> (Smylie, Olding, and Ziegler *Sharing what we know about living a good life*).

In this last section, we suggest ways that educational policies, curriculum, and teaching methods can bring forth a generation of people who embrace the kind of "original instructions" we have outlined above. Essentially, we assert that critical awareness about humanity's place in the ecosystem and employment of Indigenous worldview should come about via (a) tapping into our ancestral knowledge; (b) international partnerships already moving in this direction; adoption of UNESCO curricula; (c) federalization of schooling; and (d) more focused expectations from accrediting bodies. The problem of anthropocentrism is best served by bringing Indigenous wisdom to bear on schooling via such venues. The rationality of modern education diminishes the inner lives of those other-than-human sentients (Wohlleben, 2016a, 2016b), requiring us to pay close attention to our landscapes with a different set of assumptions. Thousands of years of indigenous experience teach us that animals, for example, display the same range of human emotions and behaviors as we do and can teach us how to live life. This is what makes sense: that evolution did not prescribe a unique biologically trajectory for humans and something different for other species. Rather, evolution builds from what is already available. The genetic programming of our ancient ancestors still works in us—and it is this genetic programming that can be our ultimate teacher if we should but lean forward, quiet ourselves, and listen.

Thus, to stop anthropocentrism from continuing to allow the destruction of life systems that disregard the intrinsic value of nonhuman life forms, we must re-embrace the worldview that kept us from destroying the planet for thousands of years. The key to accessing such epigenetic programming is to draw from the deep well of Indigenous knowledge still available in the world. Rapidly vanishing as noted above, we are in a race against our own runaway inventions. Stepping up to this challenge and opportunity can best approached by setting a policy framework that encourages the teaching of Indigenous knowledge, developing curricula for teacher preparation programs, and supporting teachers in the classroom. Committing to teaching traditional Indigenous know-how requires sensitivity to those who still, against all odds, possess the knowledge. This means a simultaneous commitment to protecting Indigenous rights; ending settler colonization that continues to oppress First Nation sovereignty; and accessing Indigenous sources when possible. Books such as *Teaching Truly: A Curriculum to Indigenous Mainstream Education* (Four Arrows, 2013) and *Indigenous Sustainable Wisdom: First Nation Know-how for Global Flourishing* (Narvaez and Four Arrows, 2019) can be of use but if there are local Indigenous People in places where sustainability education is focused, this is always the best option.

An Indigenous Policy Framework: A Radical Approach

In the United States, education is left to the purview of the states under the tenth amendment of the Constitution. This amendment confirms that any item that is not explicitly noted in the constitution shall fall under the authority of the states. As such, the US is among one of the only developed countries in the world that does not have a federally run education system. There is, however, a federal department of education (ED) that was created in 1980 with the passage of the Department of Education Organization Act. The mission of the Department is to "promote student achievement and preparation for global competitiveness by fostering educational excellence and ensuring equal access (ED Mission, 2018)." Over the years, a number of initiatives have been created by ED to facilitate the execution of their mission at the state level. By providing incentives for initiatives that are seen as advancing the country's global competitiveness through the education of its citizens, the federal government has a significant opportunity to promote certain ideas.

For example, ED routinely sponsors a substantial number of initiatives, often in partnership with other national associations. Just to name a few, there are specifically funded initiatives on civic learning, early learning, green strides (a program to support and honor schools that include sustainability education), innovation, literacy, rural education, and teacher preparation (U.S. Department of Education, 2018). As a first step to setting the appropriate policy framework for K-12 education, we propose two avenues for exploration: first, that principles of applied Indigenous knowledge frameworks become embedded into the existing green strides initiative, and second, that a separate initiative is created for applied Indigenous knowledge to be funded and encouraged to be taught in the nation's classrooms.

The green strides program emerged from the Green Ribbon Schools Program, which was created as a result of the advocacy of 80 nongovernmental organizations (ED Program, 2018). Originally, an award program, the green strides initiative, was added to connect all school communities with free, publicly available resources from a variety of sources. The Green Strides portal was created to be used as a one-stop shop for resources, webinars, case studies, promising practices, and collaboration so that all schools can make progress toward achieving the goals of the award. The purpose of green strides is to offer educational resources, professional learning networks, and tools in order to equip supporters to educate students about global sustainability and deepen learning by acting in their communities (Center for Green Schools, 2018). One way that they achieve this goal is by offering an inexpensive online certificate program to teachers and other interested professionals in the subject on how to create a green classroom. Topics include supporting environmentally responsible practices by saving energy, saving water, improving indoor air quality, and fostering an appreciation among future generations for environmentally sustainable practices.

These topics are a start and may be a good way to find the organizations and people most likely to support creating more diverse curricula, one that integrates the philosophy and application of Indigenous knowledge. In addition, one can imagine more inclusive curricula, in that content is taught in a holistic way rather than in the western epistemological framework of specialized, siloed knowledge. Imagine a similar resource portal for teachers with a certificate program on applied Indigenous knowledge. Objectives for such a course could be derived from UNESCO principles already established on this subject. Courses could be designed to foster participants to:

- Appreciate Indigenous perspectives on ways of living together and using resources sustainably.
- Appreciate the role of Indigenous knowledge and traditional ways of learning in maintaining the sustainability of a community.
- Understand the role of "modern" education in undermining Indigenous knowledge and ways of teaching and learning.
- Identify opportunities for integrating relevant aspects of Indigenous knowledge and approaches to teaching and learning into the school curriculum.

By providing a portal for resources and a curriculum for teachers, we can begin to integrate Indigenous knowledge frameworks within existing structures, such as in the case of the green strides program.

The above is merely a beginning. Second will be the importance of advocating for a separate initiative with the U.S. Department of Education to create a new project called *Teaching and Learning for a Sustainable Future* (UNESCO, 2018a, 2018b). The purpose would be to create a special home, supported at the federal level, that illustrates ways that Indigenous knowledge may be integrated into education and thereby bring the benefits of helping to "sustain" Indigenous knowledge and societies to all. It could also encourage teachers and students to gain enhanced respect for local culture, its wisdom, and its ethics, and provide ways of teaching and learning locally relevant knowledge and skills. Potential allies could be environmental organizations, folk schools, Indigenous communities, and national associations and networks such as the American Association for Sustainability in Higher Education (AASHE) and the Alaskan Native Knowledge Network (ANKN). Building a broadly based coalition will be an important step to infusing traditional education with the principles of Indigenous knowledge, which will be critical to humanity's future. This work will position our society to be better prepared to create the systems so desperately needed to stop our collision course with a global environmental catastrophe.

While all this is possible and even admirable at the federal level, it will be important to persuade school systems of the importance of this work. One possible way to do this is through the accrediting bodies. Accreditation in the United States is in the hands of regional accreditors, of which there are seven scattered across the country. There are also some nationally focused accrediting associations, such as the one created for distance learning. There are a few other entities, which focus on specific disciplines such as business, law, medical, and clinical professions. For the purposes of supporting Indigenous knowledge, the most influential of these bodies would be the regional accreditors. Accreditation approves institutions, not specific content for programs. However, they do set the framework and standards by which institutions are affirmed. For example, the Accrediting Commission for Schools Western Association for Schools and Colleges (ACS WASC) evaluates schools on capacity, commitment, and competence to support high-quality student learning and ongoing school improvement. The school assesses its program and its impact on student learning with respect to ACS WASC criteria (ACS WASC, 2018). These criteria could conceivably include language that would support the inclusion of Indigenous knowledge into the curricula. For example, it is common practice to support diversity and inclusion as part of the education mission of schools. If educators were prepared in Indigenous knowledge, they could make the case for broadening current epistemological frameworks into diversity and inclusion that would support student development in critical thinking, complex problem solving, and cognitive complexity.

To summarize, our proposal advocates for tackling the integration of applied Indigenous knowledge into current classroom teaching through a combination of federal support and regional accreditation processes. At the federal level, there already exist potential allies and initiatives that could be framed more broadly to include Indigenous knowledge. In addition, a portal for resources for teachers as well as programs that they could participate in and use for their ongoing professional development is the key to success. At the regional accreditation level, it will be important to make the case that Indigenous knowledge is a component of diversity and inclusion. As many scholars have pointed out, knowledge needs to be holistically understood and diverse perspectives need to be cultivated, in order to create a sustainable future for humanity.

We would also like to remind our readers that Indigenous knowledge is ancient wisdom, actively cultivated over a period of time and critical to be included in the conventional ways that we think about knowledge. A good example that illustrates the roles of each side by side can be found in the chart below that compares Indigenous and formal education:

Aspects of education	Indigenous education	Formal education
View of knowledge	• Sacred and secular together; includes the spiritual • Holistic and integrated—based on a whole systems view of knowledge • Stored orally and in cultural practices • Powerful predictability in local areas (ecological validity) • Less valued in distant areas	• Secular only; often excludes the spiritual • Analytical or reductionist—based on subsets of the whole • Stored in books and computers • Powerful predictability in natural principles (rational validity) • Weak in local use of knowledge
Objectives	• Long-term wisdom • Cultural and ecological sustainability • Practical; for use in everyday life • Integration of critical thinking and cultural values in decision making	• Short-term recall • Economic sustainability • Abstract; to pass examinations • Use of logical and critical thinking in making decisions
Methods of teaching and learning	• Lengthy period of acquisition • Learning through experience • Teaching through example, modeling, rituals, and storytelling • Tested in practical life situations	• Rapid acquisition • Learning by formal education • Teaching through abstract concepts and didactic methods • Tested artificially in examinations

This chart (UNESCO, 2018a, 2018b) can serve as an ongoing guide to teachers so that they can hold both sets of knowledge as they create their classrooms and inspire others how they can cultivate their own knowledge and wisdom, based on the world around them. While these ideas may be ambitious, it will take no less to re-discover the genetic programming of our ancient ancestors that is still within each one of us.

References

ACS WASC. (2018). Accreditation process overview. Retrieved form https://www.acswasc.org/schools/acs-wasc-school-process-overview/

Bagchi, A. (1996). Conflicting nationalisms: The voice of the subaltern in Mahasweta Devi's Bashai Tudu. *Tulsa Studies in Women's Literature, 15*(1), 41–50.

Balcombe, J. (2011). *Second nature: The inner lives of animals*. London: St. Martin's Griffin.

Balcombe, J. (2017). *What a fish knows: The inner lives of our underwater cousins*. New York, NY: Farrar, Straus and Giroux.

Cajete, G. (1994). *Look to the mountain: An ecology of indigenous education*. Skyland, NC: Kivaki Press.

Center for Green Schools. (2018). *What we do*. Retrieved from http://centerforgreenschools.org/what-we-do

Chomsky, N. (2013). *Cover endorsement for Four Arrows. Teaching truly: A curriculum to indigenize mainstream education*. Bern: Peter Lang.

Clarke, J.-A. (2015). The integrative entrepreneur: A lifeworld study of women sustainability entrepreneurs. In D. B. Willis, F. Steier, & P. Stillman (Eds.), *Modern strategies for sustainability leadership* (Fielding Graduate University monographs) (Vol. 5, pp. 202–233). Santa Barbara, CA: Fielding Graduate University.

Coomaraswamy, A. (1918). *The dance of Shiva*. Retrieved from https://babel.hathitrust.org/cgi/pt?id=nyp.33433082246665;view=1up;seq=10

Coppo, L. (2004). *The color of freedom: Overcoming colonialism and multinationals in India*. Monroe, LA: Common Courage Press.

ED Mission. (2018). *Mission of the U.S. Department of Education*. Retrieved from https://www2.ed.gov/about/overview/mission/mission.html

ED Program. (2018). *U.S. Department of Education Green Ribbon School*. Retrieved from https://www2.ed.gov/programs/green-ribbon-schools/index.html

Ereira, A. (2012). *Aluna (conscience)/the Kogi*. Retrieved from www.alunathemovie.com/the-kogi/

Four Arrows (2016) Point of Departure: Returning to Our Authentic Worldview for Education and Survival. Charlotte, NC: Information Age Publishing.

Four Arrows (2013). Teaching Truly: A Curriculum to Indigenize Mainstream Education. New York: Peter Lang.

Goettner-Abendroth, H. (Ed.). (2009). *Societies of peace: Matriarchies past, present and future*. Toronto, ON: Inanna Publications.

Goettner-Abendroth, H. (2018). Re-thinking 'matriarchy' in modern matriarchal studies using two examples: The Khasi and the Mosuo. *Asian Journal of Women's Studies, 24*(1), 3–27.

Gupta, S. M. (1971). *Plant myths and traditions in India*. Leiden: Brill.

Kimmerer, R. W. (2013). *Braiding sweetgrass: Indigenous wisdom, scientific knowledge and the teachings of plants*. Minneapolis, MN: Milkweed Editions.

LeGrand, K. J. (2015). Beyond conservation: Exploring the values and norms of environmental activists. In D. B. Willis, F. Steier, & P. Stillman (Eds.), *Modern strategies for sustainability leadership* (Fielding Graduate University monographs) (Vol. 5, pp. 50–92). Santa Barbara, CA: Fielding Graduate University.

Lovejoy, A. O. (1933/2010). *The great chain of Being*. https://books.google.com/books?id=ByHNG8GzUeAC&printsec=frontcover&source=gbs_ViewAPI#v=onepage&q&f=false

Maathai, W. (2003). *The green belt movement: Sharing the approach and the experience*. New York, NY: Lantern Books.

Mann, B. A. (2016). *Spirits of blood, spirits of breath: The twinned cosmos of indigenous America*. Oxford: Oxford University Press.

Matthiessen, P. (2008). *The snow leopard* (Vol. 1978). Park Imperial, NY: Penguin.

Mills, A. and Slobodin, R. (2015). Amerindian rebirth: Reincarnation belief among North American Indians and Inuit. Toronto: University of Toronto Press.

MacGillivray, A. (2015). Leadership, Boundaries and Sustainability. In D. B. Willis et al. (Eds.) Sustainability Leadership: Integrating Values, Meaning, and Action: Volume 5 (Fielding Monograph Series). Santa Barbara: Fielding Graduate School Press.

Narvaez, D., Four Arrows, Halton, E., Collier, B., and Enderle, G. (Eds.) (2019). Indigenous Sustainable Wisdom: First Nation Know-how for Global Flourishing. New York: Peter Lang.

Obeyesekere, G. (2002). *Imagining karma: Ethical transformation in Amerindian, Buddhist and Greek Rebirth*. Berkeley, CA: University of California Press. (See also http://www.ahandfuloflleaves.org/documents/Imagining%20Karma_Obeyesekere.pdf)

Olson, R. (1949). *Field notes taken at Rivers Inlet and Bella Bella, British Columbia*. Berkeley, CA: Bancroft Library, University of California at Berkeley. Referenced in Harkin, Michael (2013) "Person, Time and Being" (p. 208) in Mills, Anonia and Slobodin, Richard, *Amerindian Rebirth*. University of Toronto.

Reddy, J. (2013, Ootober). What Colombia's Kogi people can teach us about the environment: The Kogi people are warning society of destruction we face if we fail to embrace nature. *The Guardian, 29.* Retrieved from https://www.theguardian.com/sustainable-business/colombia-kogi-environment-destruction

Right Livelihood Award. (2018). Retrieved from http://www.rightlivelihoodaward.org

Schein, S. (2015). Cultivating an eco-psychological Foundation for Deep Sustainability Leadership. In D. B. Willis, F. Steier, & P. Stillman (Eds.), *Modern strategies for sustainability leadership* (Fielding Graduate University monographs) (Vol. 5, pp. 166–201). Santa Barbara, CA: Fielding Graduate University.

Smith, B. G. (2008). *The Oxford encyclopedia of women in world history* (Vol. 1). Oxford: Oxford University Press.

Smylie, J., Olding, M., & Ziegler, C. (2014). Sharing what we know about living a good life: Indigenous approaches to knowledge translation. *JCHLA/JABSC, 35*, 16–23. https://doi.org/10.5596/c14-009. Retrieved from http://www.wellivinghouse.com/wp-content/uploads/2014/04/sharing-what-we-know.pdf 2014

Stillman, P. (2015). Sustainability as organizational culture: Uncovering values, practices, and processes. In D. B. Willis, F. Steier, & P. Stillman (Eds.), *Modern strategies for sustainability leadership* (Fielding Graduate University monographs) (Vol. 5, pp. 13–49). Santa Barbara, CA: Fielding Graduate University.

Tudge, C. (2006). *The secret life of trees: How they live and why they matter.* Park Imperial, NY: Penguin.

U. S. Department of Education. (2018). *ED initiatives.* Retrieved from https://www2.ed.gov/about/inits/ed/index.html

UNESCO. (2018a). *Indigenous and formal education.* Retrieved from http://www.unesco.org/education/tlsf/mods/theme_c/mod11.html?panel=5#top

UNESCO. (2018b). *Indigenous knowledge and sustainability.* Retrieved from http://www.unesco.org/education/tlsf/mods/theme_c/mod11.html

Webb, H. S. (2012). *Yanantin and Masintin in the Andean world: Complementary dualism in modern Peru.* Albuquerque, NM: University of New Mexico Press.

Welcome to the Anthropocene. (2011, May 26). *The Economist.* Retrieved from www.economist.com/node/18744401

Welcome to the Anthropocene. (2013). *Welcome.* Retrieved from www.anthropocene.info/

Welcome to the Anthropocene on Vimeo. (2012, May 23). A film about the state of the planet. Retrieved from https://vimeo.com/39048998.

White, D. M., & Guyette, S. M. (2010). *Zen birding.* Ropley Hants: John Hunt Publishing.

Willis, D. B., Steier, F., & Stillman, P. (Eds.). (2015). *Modern strategies for sustainability leadership* (Fielding Graduate University monographs) (Vol. 5). Santa Barbara, CA: Fielding Graduate University.

Wohlleben, P. (2016a). *The hidden life of trees: What they feel, how they communicate—discoveries from a secret world.* Vancouver, BC: Greystone Books.

Wohlleben, P. (2016b). *The inner life of animals.* New York, NY: Vintage Books.

Chapter 8
Relational Conscientization Through Indigenous Elder Praxis: Renewing, Restoring, and Re-storying

Amanda Holmes and Sara Tolbert

> *We need acts of restoration, not only for polluted waters and degraded lands, but also for our relationship to the world. We need to restore honor to the way we live, so that when we walk through the world we don't have to avert our eyes with shame, so that we can hold our heads up high and receive the respectful acknowledgment of the rest of the earth's beings — Robin Kimmerer*

> *We must first detail what we value about intelligence to even see there are other interpretations of life, brilliance, and knowing — Manulani Meyer*

Elder Praxis: Land, Language, and Relations of Spirit

The land is alive, populated, and storied by spirit—and she is known as a mother to those whose memories are long enough, expansive and flexible enough, to be able to carry this knowledge, this quality of relatedness. Relations of knowledge, of knowing. This way of knowing requires longevity of relational memory and relational experience within places—lands and waters—that birth epistemologies vast enough to hold on to the fundamental, orienting reality that all beings (including human beings, located within the being-ness of the universe, as but one tiny part) are inextricably connected with each other in profound, unknown ways, ways grasped through understandings of spirit, as intricately, intimately relational.

A. Holmes (✉)
Amanda Holmes, Independent Scholar, Tucson, AZ, USA
e-mail: tokalasa@gmail.com

S. Tolbert
Te Whare Wānanga o Waitaha University of Canterbury, Christchurch, New Zealand
e-mail: sara.tolbert@canterbury.ac.nz

A. S. Gkiolmas, C. D. Skordoulis (eds.), *Towards Critical Environmental Education*, Critical Studies of Education 14,
https://doi.org/10.1007/978-3-030-50609-4_8

Lands and her languages recognize, speak, listen to the language of her children, whether those children are rivers or animals, plants, stones, or thunder, each speaking their own language in and to their own places and each other, an undulating, spiraling constellation of sensory-spiritual language and listening. Onkwehon:we, the Real Human Beings, still recognize—and participate in—these ancient, continuing languages of relationship within their lands, and continue to try to uphold these spinning worlds of their ancestral relationships in their homelands, even while these worlds of relations are being strangled, suffocated by those who refuse to wake up.

Indigenous Elders awaken relational memory, enacting, engaging, and embodying knowledge that comes from within the interaction and exchange of the generations with each other, our relations with the rest of Creation, to the land as alive with relatives, with story, with language and ethical protocols of activity, ancient systems of knowledge that understand and perceive how to *be* in the places and with the relations they have been gifted to live with and respect and take care of, a sacred compact with their Creator: intergenerational wisdom practices and knowledge relations. Onkwehon:we Elders hold and carry close to themselves ways passed down to *them*, a generosity of the generations, reaching backward and forward and into the center all at once, spiraling coherently into the past and the future and all permutations of the now, since the beginning of time. Elders embody this intergenerational generosity, a process of ethical reciprocity and a cycling of ways of knowing-being that can be conceptualized in Maori scholar Makere Stewart-Harawira's (2005) articulation of the spiral as,

> a hermeneutic framework that locates indigenous cosmologies and ways of being at the centre of an expanding spiral of being…Here the metaphor of the spiral signifies the turning back 'on a wheel of strength,' to 'the place it came from': in other words, to the sacred teachings of the ancestors, to the source of 'the primal energy of potential being', and the returning of these to the forefront in a dynamic process of re-creation and transformation (p. 24).

This Elder Praxis reinvigorates our relationships to and our memory of land as alive, as storied, as spirited, containing and constantly creating and re-creating the essence and lifeblood of form and presence, beingness, remembering and reminding us of our relatedness—and the ways of honoring and giving back to Creation with our very lives and bodies, the ways we choose to walk. Indigenous knowledge and language have become embedded within the land over millennia. Elders remind us of this by *living* their ways of knowing within their landscapes of home; through their process of lived coherence, they invigorate a vast web of interconnected fibers that hold on to the connections between us and our relatives, our collective, synergistic ways of being and knowing. The embodied praxis of Elders awakens, renews, and restores the relationships between human beings and Creation.

Moving Beyond Settler-Colonial Thinking

Theorizing, engaging, and enacting Indigenous intergenerational relationality and reciprocity creates conditions for reimagining resurgence. Intergenerational resurgence is impossible without the reestablishment of intergenerational protocols, practices, and relationships that are deeply contextual, but have been disrupted. As part of the Settler-Colonial project, Elders, Knowledge Holders, medicine people, and children were strategically targeted for eradication and removal from the community so as to disrupt and destroy cultural coherence and intergenerational connections of knowledge, connections that constitute power. In disrupting and destroying roles, relationships, and connections with each other, the colonizers knew they were simultaneously disrupting and destroying Indigenous relationship with and connection to their lands. Reconnecting to these relations of lands, natural world, languages, and each other through awakened intergenerational interaction and exchange lies at the heart of Indigenous resurgence. From Onkwehon:we perspectives and protocols, Elders and their ways of knowing-being are the center of resistance, as resistance means continuing, persisting, maintaining, creating, remembering, visioning, and dreaming *beyond* Settler realities.

Our resistances must evolve, transform, shift, and renew themselves outside of Settler-Colonizing thinking, beyond the limitations of its capacity to conceptualize, beyond its asphyxiating epistemologies of dominion. They must ground and inspire and re-center, creating from within places and spaces that do not inhabit the same conceptual universe—because they are not the same. Settler-Colonialism may be consuming the world as we know it (as it always has, only now in ever more mutating, deadly, and sophisticated shapes and forms), it may be attempting to control and consume all definitions and meanings and knowings, and as much as we may be surrounded by it, we do not have to allow ourselves to become *of* it (Grande, 2018). Free of the numbing cooptation of *homo assimilans* (Wolfe, 2016), our voices and lives and languages have the capacity to engage and participate with all those we *know* to be our relatives.

We need so desperately now to join our resistances—this resistance from a different center, a multitude of centers—to those of our relations, *who are also resisting*, who are also trying to awaken our remembering of older knowledges of our relationships. Ocean and Deer and Mountain and Pine, who know the settler well, who recognize its energy, have never relinquished their Original Instructions to be who they are, because *their knowledge of who they are* has been given to them by the original energy of the universe and of creation. It is, still, their essence and their knowledge and their way of being. And they too are our models of resistance, of how to live surrounded but never give up who we are and what that means.

In scholarship and dominant discourse around environmentalism, education, and environmental education, the Western scientific monopoly on Truth and its relationship to power are implicit, as it is assumed that Western scientific approaches and epistemologies are the singular path to the singular Truth, the only way to knowledge that counts (Battiste, 2000, 2011; Brayboy, Gough, Leonard, Roehl II, &

Solyom, 2012; Smith G.H., 2000; Smith L.T., 1999). Indigenous perspectives are marginalized and excluded, not considered relevant by dominant Whitestream educators and academics, unless as an add-on to their paradigm of Truth.

> Western scientific method has historically been presented as neutral, objective, and representative of the Truth. Research grounded in these methods has functionally served to vivisect the world, cutting across interconnections, lives, cultural knowledge, and bodies, often with good intentions and *occasionally espousing a critical approach even as it reproduces the status quo* (emphasis added, Brayboy et al., 2012, p. 428–9).
>
> For example, Western methodologies often assume the power to define taxonomically what is human or nonhuman, animate or inanimate, organic or inorganic, living or lifeless, natural or unnatural, rational or irrational. In addition to promoting rigid definitions and labels, Western scientific methodologies may seek to exclude other epistemologies and methodologies that focus on the processes and qualities of relationships between and among humans and the worlds they inhabit (Brayboy et al., 2012, p. 432).

Indigenous Elders are thrown even further to the peripheries by Western ways of knowing—while fetishized, their knowledge, their ways of knowing-being, their ways of communicating are made marginal, subverted and undermined (and thus avoided), as less-than, irrelevant, anachronistic. The deep value placed on Elders and their role in Indigenous societies as the heart of understanding, of awakened consciousness, knowledge, and wisdom illuminates an epistemological chasm, and the enormous difficulty in communicating across vastly divergent epistemological universes.

On Elders and Elder Praxis

For non-Indigenous people, there is a certain incomprehension about the meaning of the term "*Elder*," so it is important to clarify what "Elder" means from Indigenous vantage points. In the U.S., "the elderly" are pushed to the margins, having lost their value and usefulness in a society hyperfocused on superficiality, youth, speed, fad, and materialism. The marginalization of Elders is also embodied in critical or progressive education movements that center youth discourses and practices without attending to those of Elders in the community. The United States consumes its elderly; "care" for the elderly is often sterile, medicalized, and underfunded. For Indigenous Peoples, Elders hold the deep respect, reciprocity, and reverence that form the heart center of their People, having lived and survived through a lifetime of experience, gaining the recognition of their community as one of the most esteemed roles and positions in Indigenous societies. Elders possess knowledge gathered out of lifelong experience and participation, their praxis nurtured, guided, and sustained by the praxis of their own Elders in a cycling of intergenerational reciprocity. These are cycles of collective community coherence that are maintained, developed, and transformed through each generation by the careful, deliberate, perceptive ways of knowing-being of Indigenous Elders. Elders do not only carry "the knowledge," per se, but it is in *the ways* they live their knowings, the ways they enact and animate Indigenous knowledge systems, that these knowledges come to life from within the spiraling of the generations. Cree Elder Joe Couture refers to this when he says, "True Elders are superb embodiments of highly developed human potential" (Couture, 1991, p. 207–8). As Lakota Elder and Buffalo protector Rosalie Little Thunder always emphasized, "Elders are the center. Because Elders are the mirrors of our Ancestors" (Little Thunder, personal communications). Living, oral ancestral collective knowledge systems are a sustaining power of Indigenous communities, and these are carried and continued by Indigenous Elders and Knowledge Keepers.

Elder Praxis is a way to think more deeply about Onkwehon:we Elders' ways of knowing and the intergenerational, inter-relational links of protocol and survivance of the human and other-than-human beings, where Elders hold a vital place in the web of "knowledge production and exchange"(Meyer, 2013, p. 98), situated at the nexus of the generations. Elder Praxis is walked in the everyday by Onkwehon:we Elders, a relationality of thinking and living that is at once individual and collective, ancestral and contemporary and of the future, responsive, diverse, particular, expansive, processual, responsible, reciprocal, generous, ethical, just, self-determining, and transformative.

Growing up in her Lakota oral processes and protocols, Rosalie Little Thunder talked about experiencing the uniquely important social, cultural, spiritual space held by Elders, as closest to Ancestors and keepers of future generations (Little Thunder, personal communications). Onkwehon:we Elders form a genealogy of collective knowing from within their ancient home places on the back of this Great Turtle, a longevity of knowledge relations that engages their grandparents, their grandparents' grandparents, and on and on, in ongoing interaction with generations of human and other-than-human beings, in their places of homeland and collective narrative memory (McLeod, 2007); that is to say, remembering, from deep within context.

In the everyday, embodied praxis of Elders *living* their ways of knowing-being, their ethical knowledges and worldviews, from within contextual, intricate patterns of relationship between language, knowledge, and land, presences and continuities, traplines of memory and knowing, the awakening of intergenerational consciousness and practice, and the re-imagining of community, is restored and renewed. An intricately woven web links us to each other, to Spirit and the natural world, to Earth, to embodied community, to Ancestors, to teachings and philosophies and ceremony, the re-generation and re-envisioning of ancient memory into the everyday of today within the continuity of intergenerational oral practice. Restoring these relational ways of knowing and being, storying and re-storying these connections, remembering the cyclical nature of their patterns and the particular constellation of beings, energies, elements, and relationships from deep within their context, is the *how* of renewal and resurgence, the practices of wisdom carried within ancient philosophies and teachings, alive today within the praxis of Onkwehon:we Elders, the generations, and our still-living natural world.

Elders' lived reflection, *reflexion*, of/on their own Ancestors' ways of knowing and being *are* the teachings, embodied in the walking of their lives, their approaches to living, the choices they make and how they make them—these are the cultural teachings, the knowing, and the wisdom themselves. Elders both carry and embody wisdom—the collective, lived ethic of their People—as intergenerational praxis. Indigenous common sense. With time, energy, commitment, discernment, and the development of our perception, we learn to attune ourselves to what they choose to share with us; and *what*, in Indigenous knowing, is no noun, it is active, as *ways* of doing-knowing-being (Betty Carr-Braint, personal communications). Going deeper, we align ourselves within this way of active/activated Onkwehon:we orality, older ways of listening, that pay close attention to *how* our Elders are walking their paths through their lives, *why* they choose to do it that way, and what that *means*. Nurturing intergenerational renewal and "maintain[ing] those cycles of continuous creation" (Cook, 2008, p. 165), we learn how this active, activated perception radically, at the roots, comes to shift the way we live our lives in return, the choices we encounter and the coherence of clear-mindedness with which we choose, moment to moment, that make up the heart and spirit of our own walk. Onkwehon:we reciprocities of active, transformative relationship emergent from within ancient, ancestral practices of the generations…unfolding into and through us in the now. And the here.

> *Be grateful that somebody's looking after that spiritual side of things. That's what's the real world, actually, the spiritual side of it, and you can't see [it], but it's probably the most real of all. So we have to bring the rest of the world into that context. They have to understand the relationship and the responsibilities. — Oren Lyons*

Re-imagining Human/Beyond-Human Collectivities and Solidarities of Resistance and Resurgence: Renewing, Restoring, Re-storying

This is what resistance *also* looks like—Indigenous peoples reconnecting within their *own* frameworks of knowledge and the generations, lands, memory, and oral land-based practices, language, relatives, and the praxis of Elders as the heart center of awakened presence-ing to each other, a resurgence of Onkwehon:we original consciousness and coherence, remembering and comprehending collectivity and solidarity with the Beyond-Human Beings, the realm of the "seen and unseen." Onkwehon:we knowledges understand that human beings and beyond-human beings are embedded within an interlocking relationship of resistance and survivance that the human beings, given our weaknesses and limitations, are only beginning to (re)awaken to. Paying attention now to the intertwining resistances of the human and beyond-human beings will require thinking in a nonlinear, nonpositivistic way. It is time to do what Manulani Meyer suggests, to engage "a different sensory immersion, a heightened sense of context, and a whole different tool belt useful in shaping cultural priorities for different understandings" (Meyer, 2013, p. 99). It is time—no, it is way past time—to hear, to radically listen to, the expansive, dynamic, and dialectical relationality of Indigenous epistemologies and that which Indigenous epistemological thinking allows, the possibilities it envisions.

Beyond-Human-Beings' resistance reflects at once *their own equally critical struggle for survivance*, "an active sense of presence, the continuance of native stories" (Vizenor, 1999, p. vii) (we envision "native" here as including Indigenous Peoples *and* the Beyond-Human Beings of this land), while also reflecting an active state of awareness and engagement maintained to awaken their human relatives, recognizing as they do the interlocking fate of our common destinies. Survivance applies equally to the Beyond-Human, as Beings, as alive within consciousness and spirit, from within Indigenous epistemologies. In a conscientization process that asks us to comprehend the ways that the survivance and resistance of Human Beings and Beyond-Human Beings intertwines, and in the responsibility of articulating the existence and integrity of storying that reaches the Beyond-Human universe, *survivance* might become, within this intertwining world, "an active sense of presence, the continuance of [*their own*] stories" (emphasis added, Vizenor, 1999, p. vii).

Resistance of the Beyond-Human Beings can be found in the way Spotted Owls refuse to surrender their offspring to clear-cutters; the way Deer, Elephants, Dolphins mourn their dead, staying by their side in the face of human destruction and indifference; the way Salmon attempt to return to their traditional places of spawning and breeding, maintaining their migratory paths no matter the devastation and interference; the way Grey Whales beach themselves, giving themselves up to sand and air. Through all this, they are steadfast in their persistence and continuance, maintaining their commitment to upholding their compact with their Creator. They refuse to quietly acquiesce to the dismal trajectory of the settler-colonizer-capitalist death march, even as they are forced to encounter and cope with them from within their own ways of knowing, being, and understanding. They persist as well in their efforts to heal and repair—acts of love and care in resistance. We see this with the Grey Whales in the lagoons of Baja California, who come up to play with the pescader@s, tourists and guides, despite long histories of antagonistic encounters of whalers killing their calves, and mother whales destroying fishing boats in acts of vengeful pain and agony. All of this occurred before the "proverbial peace treaty" of the 1970s when Grey Whales reached out to the boating humans, an intervention that strives to fundamentally change the nature of those relationships (Peschak, 2017), enacting a radical conscientization, resistance, and transformative praxis. Grey Whales are our Elders, repairing, rebuilding, re-storying.

Reflecting our Relatives, it is the same with Original Peoples, the Real Peoples. We recognize the fact of being surrounded—but through the ancestral coherence of generations sustained by the praxis of Elders and Knowledge Keepers, we know that our resistance means to continue, to create, to renew, to story and to re-story, in ancient ways made relevant within each generation. This is a critical, deep relationality, a spiritual orientation that is foundational for Indigenous peoples from within their lands and worldviews, an orientation that recognizes a profound connection to the natural world *beyond* what is commonly considered from within Western discourse as "connection," and, fundamentally, looks to these relationships of reciprocity, mutuality, and resilience with these beings and energies as central to knowing *who we are* as Original Human Beings.

The closer we become to the other-than-human world, the closer we approach our own humanity and what it means to be human. This is not paradoxical for Indigenous people—it is part of the natural order of things, the natural law of the universe. Onondaga Faithkeeper Oren Lyons reminds us of this fundamental Haudenosaunee orientation:

> Another of the Natural laws is that all life is equal. That's our philosophy. You have to respect life—all life, not just your own. The key word is 'respect.' Unless you respect the earth, you destroy it. Unless you respect all life as much as your own life, you become a destroyer, a murderer….In our way of life, in our government, with every decision we make, we always keep in mind the Seventh Generation to come (Oren Lyons in Wall & Arden, 1990, p. 67–8).

Within such a worldview, reinforced by teachings, stories, language, song, art, ceremony, there is no epistemological challenge or cognitive dissonance to knowing the place of humans in the universe, as learners, and Heron, Cedar, Stone as the greatest of teachers:

> To learn, you have to have teachers. And who's your teacher, the teacher is nature, the earth. You learn. You learn. You learn how to get along. You learn how to be respectful….The Indigenous people have about the best understanding of this. And I would say, that's probably the biggest loss that I see in humanity now is its loss of understanding of relationship. *They don't understand the relationship* (Oren Lyons, 2016a, 16 Sept, 1:09).

Relational consciousness and knowing emerge from within Onkwehon:we epistemologies of interconnectedness, engaging responsible, respectful, thankful ways of knowing all of Life and the meanings of being a clear-minded human being, a relative. These responsibilities and commitments to *relational* ethical knowledge and value systems embedded within language, traditional teachings, and worldviews lie at the heart of Indigenous epistemologies, ways of being, pedagogies, and methodologies. They form profoundly relational epistemologies, lived and living, enacted and embodied, that consider and know the "unseen," Beyond-Human-Being world as inherent within one's self, within Onkwehon:we collectivity.

Sandy Grande asks the question, "What kinds of solidarities can be developed among groups with a shared commitment to working beyond the imperatives of Settler-Colonialism?" (Grande, 2018), while asserting that Settler-Colonial logic is embedded within and constitutive of the nature of place. A deeper knowing suggests, calls us to remember, that Settler-Colonial logics of elimination can never truly be constitutive of the nature of place, and this thinking will require a remembering that is nonmaterialistic, nonmechanistic, nonlinear, noninstrumental, a remembering of the way the relationship between Land and her Original People are alive and of spirit and being, and thus capable of escaping predatory logics, definitions, and imaginations. So then, what kinds of solidarities might be cultivated to work beyond the confines of the Settler-Colonial imaginary? What would these be based within? Where would they unfold, and how? Where do we locate a shared ethic of relationality, from which might emerge our common, collective resistances? Where does an ethic, a "logic," a consciousness of Earth as Mother, as living relative and lifegiver, reside? Glen Coulthard reminds us that, "Reinstating indigenous social relations of authority and power over those territories that are being expropriated and devoured by extractivist industries and in relation to settler communities is crucial. It requires a solidarity that's organized around indigenous peoples' relational understandings of land, autonomy, and sovereignty" (Coulthard in Epstein, 2015, par. 59). Onkwehon:we know these ethics as *who they are*, stretching through space and time, in Onkwehon:we directions, the Real People finding and enacting this ethic embedded within Onkwawen:na and deep pools of perception and practice, in ways that are most fundamentally summoned and animated by Elders.

Creating Decolonial Space(s) (Holmes, 2018), of lands and waters, memory, story, relationship, and spirit, where an abundant diversity of different, older conversations unfold, where we remember and reinvigorate different solidarities and alliances than those we have previously imagined, in ways we have yet to envision, our collective relationality as children of our living Earth might emerge. These are the kinds of spaces where we engage in redefining resistance, and solidarity, the rhizome-ing of an "expansive resistance approach" (Smith G.H., 2000, p. 69), one that challenges all of us to remember who we are by remembering our connection to all that is, all that exists, as our most basic, elemental knowing of ourselves. These practices are *alive* for Indigenous peoples within cultural ways of knowing and being, within languages, in ways that do not need to be explained or validated. For Haudenosaunee, the Thanksgiving Address, as but one example, is the practice of an animating generosity and reciprocity, as relational protocol, that becomes embodied within us as individuals and us as collective when we offer our thanksgiving to all of life, spoken through Onkwawen:na and our clear mind, within a genealogy of intergenerational longevity of relations, to our universe in our places of home.

Indigenous relational epistemologies of deep reciprocity, generosity, humility, respect, and thanksgiving lie at the root of the need to restore and renew Indigenous connections to Indigenous lands that have been ruptured through Settler-Colonial dispossession (Kahsto'sera'a, Karahkwinéhtha, Karennanoron, Karonhyaken:re, Katsitsi:io, Tayohseron:tye, and Whitney at Kanatsiohareke summer language program, 2016, personal communications). Indigenous practices of intergenerational, relational consciousness and participation of the human, the seen, and the unseen worlds, require their places of interaction and exchange, wherein co-perception and co-creation become possible. This relational space needs to be held and engaged by the Original Peoples, for the benefit of all life, to keep the World in balance, as so many Indigenous Elders have articulated for so long. Cultivating solidarities with each other and with our living world—the reciprocity of spiritual, relational solidarities giving rise to spiritual activism and spiritual activism giving rise to spiritual, relational solidarities—we are challenged to "reimagine ourselves beyond our skins" (Christian & Wong, 2017, p. 7). To reimagine our very skins as part of the layered, porous, breathing membrane of Land, Water, and Sky, Creation, a fluid world of relations.

Spiritual activism, spiritual solidarities, are acts of love and healing. Remembering "a much longer now, a much wider here" (Christian & Wong, 2017, p. 11), we locate again a presence-ing of the generations to each other that becomes transformative. As Janine McLeod states in Christian and Wong's (2017) *downstream: reimagining water*, "It seems to me that transformative collective action is unlikely to emerge from our stories of past abundance unless we conceive of our love for the world in multi-generational terms" (p. 11). Kanien'keha:ka scholar Taiaiake Alfred, reflecting McLeod, calls us to remember "who we are and where we are headed," re-centering "the crucial connections that generate the sense of community—love—that is needed to overcome the disconnection and division and mutual hatreds that reinforce colonialism" (Alfred, 2004, p. 91). Alfred maintains that confronting colonialism at the root means challenging it as "a psychological state and a pattern of thinking and a way of seeing the world and other people … the absolute destruction of the Indigenous way of thinking and living" (Alfred, 2004, p. 91). Okanagan writer, scholar, and activist Jeannette Armstrong gives voice to these profound relationships, continuing to deepen our coherence around the layering interconnections of ourselves with Creation: "It is how the land communes its spirit to you: it heals people and it does this in an incredibly profound way. We need to think about how we can do more of that" (Armstrong, 2008, p. 74). Always, always making it a living practice, a way of being, a consciousness that *becomes* Indigenous transformative action.

Black critical feminist bell hooks encourages us, "We are born and have our being in a place of memory…We know ourselves through the art and act of remembering" (hooks, 2009, p. 5), and we are called (back) into the circle, to remember and immerse ourselves within our stories of Creation. Or for (un)settlers who may have lost theirs, to locate older stories that ignite new imaginaries where human beings are not removed from gardens, where there is no original split from Creation that then becomes known outside ourselves as "nature," no civilizing self-righteousness or need for a Truth that would separate us, as individuals, from ourselves and our beings, from our collectivities, humans who have become split off from the natural order of life. Epistemologies of the Real People cannot conceive of being distinct from the universe of life, places where there is no language to describe separation from "nature," from self and existence. The late beloved Onondaga Clan Mother and Elder, now Ancestor, Audrey Shenandoah speaks to the heart of this Haudenosaunee philosophy when she says,

> We have much to learn from the incredible knowledge of our ancestors which was gained long before reading and writing came about. From time immemorial, every bit of their intelligence and senses were used. Humans knew and felt relationship to all that lived and moved. Somehow that relationship must be regained. We are faced with crucial times. Changes need to be made beginning now, for our life-support system is being severely abused and mismanaged. A healthy human mind respects the gifts of life—all nature gives life. There is no word for 'nature' in my language. Nature, in English, seems to refer to that which is separate from human beings. It is a distinction we don't recognize. The closest words to the idea of 'nature' translate to refer to things which support life (Audrey Shenandoah in Wall & Arden, 1990, p. 26).

The epistemic violence of Settler-Colonialism is a symptom of a deeply diseased epistemology, one that forms an organizing isolationist logic and a strategy of control through human disconnection from, and dismemberment of, the natural world. Glen Coulthard (2013) powerfully asserts that, "For Indigenous nations to live, capitalism must die. And for capitalism to die, we must actively participate in the construction of Indigenous alternatives to it" (par. 15). Indigenous alternatives to epistemological violence create and support solidarities of the human and beyond-human worlds, a spiritual activism that comprehends the collective necessity of Indigenous (re)connection with and within lands and practices, languages, philosophies, teachings, ways of knowing and being, restoring and renewing relationalities from deep within places, contexts, histories, and the narrative memories of the Land and her People. Time and again I have heard the generosity of Indigenous Elders when they say that every human being carries the potential of waking up to these original instructions of natural law, because everyone is a child of Earth—it is just a matter of distance from these fundamental teachings, and Indigenous Peoples are closer to this knowledge and have continued to develop this knowledge over the long term (John Mohawk, 1990; Tom Porter, personal communications and 2008; Jake Swamp, personal communications). (Un)settlers access these instructions by radically reorienting themselves to the protocols of Earth, and by profoundly taking responsibility for transforming their relationships with the Original Peoples, seen and unseen.

The Highest Form of Politics Is Spiritual Consciousness: Politics as Spiritual Consciousness, Resistance as Remembering, and Radical Humility

Resistance, then, becomes not merely an organizing *against* something—it is resistance *as remembering*, resistance *as (re)connection*, resistance *as relationship*, "revolutionary," transformative, at its most basic and most expansive layer, capable of breaking open our stuck places where we have become numb within an epistemology of deep separation. We need to call upon "our own radical remembering of our future" (Meyer, 2003, p. 54), where something that is both far beyond ourselves and deeply within ourselves calls us to wake up to our relational place in the natural order of life. We return to the need for perceiving the callings and connections of spirit that make sense of what John Mohawk means when he says, "The highest form of politics is spiritual consciousness" (Mohawk, 1990, 9:54); to be governed, guided in our everyday, by an internal ethic and engagement of Spirit, an ancient, active, ethical framework of relationality that holds the power to wake us from consumption by a capitalist-settler colonialist mining project that is mining our very spirits (Trudell, 2001, 1:26). An ethic of relational, spirit-centered connectedness, consciousness, and transformative, insurgent/resurgent activity: intergenerational spiritual activism (Rosalie Little Thunder, personal communications) with the rest of life that comprehends your place within creation. Is it no wonder that Westerners have struggled so vociferously with Mohawk's assertion?

A first step—a giant first step—for (un)settlers must be a fundamental acknowledgment of, and responsibility for, sitting with an older, wiser way of knowing and being on the land—one that knows this place since time began. Can you hold on to that without immediately having to say something? Can you listen to what that might mean? Can you allow an ancient evolved and evolving, cycling aliveness to pierce a cultural skin that has come to believe it has all the answers? Are you able to sit with this "new" knowledge in silence, and allow that it is not "new," that a vast depth of knowledges is cycling, now, in these places you have only recently begun to call home, that peoples you have been taught to be savage, pagan, heathen hold collective knowledge that has become profound wisdom—and continue to hold these evolving knowledges and relationships today? We must recognize the stunning, deadening hubris of a worldview that believes humans to be the carrier of superior intelligence, knowledge, communication, higher-level 'reasoning,' and morality, while assuming the absolute superiority and singularity of this worldview. Such astounding arrogance emerges from within an epistemology of separation, isolation, and atomization, out of which arises the basic savagery of Western settler-colonial epistemology. For Indigenous peoples, it is a shocking insanity we do not quite know what to do with—still. The magnitude of its error and disconnect, its horror, the implications of its ignorance so colossal, that it is difficult to imagine and relate to how far removed it is from Indigenous epistemologies of relationship, reciprocity, responsibility, respect (Kirkness & Barnhardt, 1991), accountability, generosity, discernment, perception, gentleness, and reverence (Archibald, 2008), relevance, and resilience (Galla, Kawai'ae'a, & Nicholas, 2014), the intricately interwoven relations of intimacy with an interconnected universe.

Kanien'keha:ka Elder Tom Porter reminds us of Rotinonhsionni understandings of the context of human beings in Creation.

> ...We the humans were the last ones to be made of all Creation. We are last. And that's why Grandma and other elders have suggested...that because we were the last of thousands and thousands of millions of creations, the Creator might have been exhausted and *tired* by the time he came to making us. And so they say, 'We think that when he put us together, he musta not exactly put one of the wires together right. And so we have a little handicap more than all other living things…' So we are sorta like the dangerous ones. We are like the baby ones that all other life needs to take care of. And look at everything else in the world. If all we humans died, everything else would keep going! So we are the ones that are the most in need of help (Porter, 2008, p. 22–3).

Coming to accept and internalize the basic reality that this entire land of Turtle Island is still home for her Original Peoples, brimming with epistemologies, philosophies, teachings, and practices of this very Earth, is a first step in the development of Settler-Colonial humility, the necessity of a transformative praxis of radical humility *and* relationship with land and the "unseen," that might allow a stillness where perhaps the Land will become ready to begin speaking again. Radical humility offers us a place and a space to begin again—to begin to comprehend our relations and connect with a consciousness of the universe as constellations of relatives, a critically needed reawakening to new theories of embodied relativity. This land has been known, and listened to/for, as a Relative, since time began. Creation Stories. Stories that tell not only "the what" and "the how," but that ask the question, as Vine Deloria, Jr. reminds us, what does it *mean* (Deloria in Deloria, Foehner, &

Scinta, 1999, p. 134). Epistemologies and knowledge systems that search for relationships, recognizing deeply that it is only in relationship that we—all—survive.

What Does It Mean to Be a Good Relative?

Lakota Elder Rosalie Little Thunder always asked the question, "What does it mean to be a good relative?" (Little Thunder, personal communications). The need for different questions….and different answers…from within different epistemologies: good relatives maintaining relationship with the rest of the universe. Fundamental, orienting Onkwehon:we principles, consciousness, ethics, knowledges, and ways of knowing-being. How to go from the emptiness of unrelatedness to a consciousness of fundamental relationality, where all the elements of life are equally significant, the bearers of sophisticated, interwoven intelligence, sentience, communication, language, knowledge, feeling, and reasoning? For Indigenous communities, access to this knowing is direct and accessible through intergenerational knowledge relations and other ways of knowing, through longevity of experience, through interaction and exchange with Elders, who are the living models of these older ways of knowing and being. For those who find themselves living as settler or unsettler on Indigenous homelands, there is a need to pay attention differently, to ask different questions, to begin to listen differently, to perceive differently, *without having all the answers*. Radical humility: to ask for and of Other scholars of knowledges that Settler-Colonialism has intentionally, strategically excluded, made marginal and illegitimate for centuries, targeting the knowledge and the knowledge holders; to not presume the role of expert; to relinquish the stranglehold on knowledge supremacy; to create and hold spaces for Other knowledges and their keepers to come forward. Holding the space with a radical listening that emerges and engages out of radical humility, being willing to be quiet and listen, waiting for what might emerge after many long generations—these are transformations that become possible when different practices, protocols, disciplines, ethics, and intergenerational consciousness and relations are enacted around knowledge.

We know, because our Elders have told us, that there would come a time when Indigenous knowledges would be needed again, knowledge about ways to live in peace and in harmony with one another, *all* of us. That time is now upon us:

> And I would say, that's probably the biggest loss that I see in humanity now is its loss of understanding of relationship. They don't understand the relationship. How do you keep this relationship, how do you polish it, how do you keep it fresh, how do you keep it shiny, how do you work with it?. You just be grateful that it's going on. That we're keeping it up. That's what's important, is that it takes place…Be grateful that *somebody's* looking after that spiritual side of things. That's what's the real world, actually—the spiritual side of it, and you can't see [it], but it's probably the most real of all. So we have to bring the rest of the world into that context. They have to understand the relationship and the responsibilities (Oren Lyons, 2016b, 9 Nov, 1:48, 4:06).

> The statement that's most important there is to understand how closely we're related to the earth. And that we're part of the earth. And we're a part of nature. Nature is us basically. And they always talk about 'environment,' like it's something over there, or it's a category or something. No, we're in the middle of it, we're a part of it, and we affect it. We survive in it. And we *are* part of the earth (Oren Lyons, 2016b, 9 Nov, 3:37).

The issue of Settler-Colonial land theft, territoriality, occupation, exploitation, and desecration must be challenged not only by Indigenous peoples but by (un)settlers as well, as Indigenous peoples must be able to access our spaces and places in order to enact responsibilities of spirit and reinvigorate ethical practices of reciprocity, for the benefit of all life. Violet Caibaiosai, one of the Great Lakes Water Walkers, says that through her walk, she had "the realization that all the teachings of our ancestors—which had been told to me about the spirit of our ancestors being within the Earth upon which we walk, for example—are so very true and very real. This spiritual element is extremely beneficial for the holistic well-being of those willing to speak with and listen to the spirit of all Creation" (Caibaiosai, 2017, p. 111). Another Water Walker, Renee Elizabeth Mzinegiizhibo-kwe Bedard says that, "because women are given these sacred prayers and songs to sing for that sacred medicine, we therefore must speak out on the water's behalf" (Bedard, 2017, p. 95). Onkwehon:we need their lands, free of Settler-Colonial control, presence, and interference—free of Settler-Colonial ways of knowing and being—to be able to restore and renew damaged relations brought about by the practices of Western capitalist-colonialist exploitation and the epistemic violence embedded within its epistemology. Destabilizing and unsettling Settler-Colonialism happens in creative Indigenous reassertion, restoration, renewal, and resurgence of our collective, diverse resiliencies and relationships from within our diverse languages and places, our connections with lands and waters, worlds seen and unseen, mediated by the praxis of Onkwehon:we Elders.

What kinds of solidarities can we cultivate and what kinds of questions do we need to be asking? As Rosalie Little Thunder constantly—and quietly—challenged us to consider deeply from within older philosophies and epistemologies, "What does it mean to be a good relative?" (Rosalie Little Thunder, personal communications), there is a deliberateness with which she did this, from deep within her Lakota language, cultural knowledge, ethics, discipline, spirit, and her positionality as an Elder. *And* she modeled her answer as process—she *lived* her activism, *spiritual activism* she called it (Little Thunder, personal communications), with her life, in her life's walk. Her understanding and highly developed consciousness of the meaning of "solidarities of spirit" with all her relations guided people to consider and reflect on the ways the rest of Creation is calling us now to wake up, to comprehend what is happening, and to act, humbly and generously, perceiving our relatives *as ourselves* (Rosalie Little Thunder, personal communications). Cultivating survivance into seven generations, before and to come, generations interacting in reciprocating relationship with the rest of the living world, is a bit of what Rosalie was showing us in the meaning of spiritual activism as living praxis. She always made a point of remembering and recognizing, in her Lakota way, the Haudenosaunee philosophy of the seventh generation. "Just think," she would say, "Imagine that. To think in terms of the seventh generation. To make decisions from that perspective. That is just so awesome" (Rosalie Little Thunder, personal communications). Elder Praxis nurturing and nudging the development of our own perception and knowing, a reciprocity of knowledge and relationship cycling through the generations.... Regeneration.

> *It takes the instruction of the Peacemaker to be of One Mind.*
> *[The Peacemaker] said, 'When you're of One Mind, the power of the Good Mind can change anything.'*
> *So we have to somehow get to that point of unity of thought and direction and effort.*
> *And if we can do that we certainly can mitigate what's going on now.*
> *And what we're concerned [about] is for the children and their grandchildren and their children.*
> *Peacemaker said that we're responsible. And that we should be always watching.* — Oren Lyons

References

Alfred, T. (2004). Warrior scholarship: Seeing the university as a ground of contention. In D. A. Mihesuah & A. C. Wilson (Eds.), *Indigenizing the academy: Transforming scholarship and empowering communities* (pp. 88–99). Lincoln, NE: University of Nebraska Press.

Archibald, J. A. (2008). *Indigenous storywork: Educating the heart, mind, body, and spirit*. Vancouver, BC: UBC Press.

Armstrong, J. C. (2008). An Okanagan worldview of society. In M. Nelson (Ed.), *Original instructions:Indigenous teachings for a sustainable future* (pp. 66–74). Rochester, VT: Bear &.

Battiste, M. (2000). *Reclaiming indigenous voice and vision*. Vancouver: UBC Press.

Battiste, M. (2011). Cognitive imperialism and decolonizing research: Modes of transformation. In C. Reilly, V. Russell, L. Chehayl, & M. McDermott (Eds.), *Surveying borders, boundaries, and contested spaces in curriculum and pedagogy* (pp. xv–xxviii). Charlotte, NC: Information Age Publishing.

Bedard, R. E. M. (2017). Keepers of the water. In D. Christian & R. Wong (Eds.), *Downstream: reimagining water* (pp. 89–106). Waterloo, ON: Wilfrid Laurier Press.

Brayboy, B. M. J., Gough, H. R., Leonard, B., Roehl II, R. F., & Solyom, J. A. (2012). Reclaiming scholarship: Critical indigenous research methodologies. In S. D. Lapan, M. T. Quartaroli, & F. J. Riemer (Eds.), *Qualitative research: An introduction to methods and designs* (pp. 423–450). San Francisco, CA: Jossey-Bass.

Caibaiosai, V. (2017). Water walk pedagogy. In D. Christian & R. Wong (Eds.), *Downstream: Reimagining water* (pp. 107–112). Waterloo, ONT: Wilfrid Laurier Press.

Christian, D., & Wong, R. (2017). *Downstream: Reimagining water*. Waterloo, ON: Wilfrid Laurier Press.

Cook, K. (2008). Powerful like a river: Reweaving the web of our lives in defense of environmental and reproductive justice. In M. K. Nelson (Ed.), *Original instructions: Indigenous teachings for a sustainable future* (pp. 154–167). Rochester, VT: Bear &.

Coulthard, G. (2013). For our nations to live, capitalism must die. *Unsettling America: Decolonization in Theory & Practice*. Retrieved from https://unsettlingamerica.wordpress.com/2013/11/05/for-our-nations-to-live-capitalism-must-die/

Couture, J. (1991). The role of native elders: Emergent issues. In J. W. Friesen (Ed.), *The cultural maze: Complex questions on native destiny in western Canada* (pp. 201–217). Calgary, AB: Detselig Enterprises.

Deloria, B., Foehner, K., & Scinta, S. (Eds.). (1999). *Spirit & reason: The Vine Deloria, Jr., reader*. Golden, CO: Fulcrum Publishers.

Epstein, A.B. (2015, January 13). The colonialism of the present: An interview with Glen Coulthard. *Jacobin Magazine Online*. Retrieved from https://jacobinmag.com/2015/01/indigenous-left-glen-coulthard-interview

Galla, C. K., Kawai'ae'a, K., & Nicholas, S. E. (2014). Carrying the torch forward: Indigenous academics building capacity through an international collaborative model. *Canadian Journal of Native Education, 37*(1), 193–217.

Grande, S. (2018, February). *Keynote address at the University of Arizona, College of Education, Teaching, Learning and Sociocultural Studies Graduate Student Colloquy*.

Holmes, A. (2018). *Geographies of home, memory, and heart: Mohawk Elder praxis, land, language, and knowledge woven in place* [Doctoral dissertation, University of Arizona]. ProQuest Dissertations and Theses Global.

hooks, b. (2009). *Belonging: A culture of place*. New York, NY: Routledge.

Kirkness, V. J., & Barnhardt, R. (1991). First nations and higher education: The four Rs—Respect, relevance, reciprocity, responsibility. *Journal of American Indian Education, 30*(3), 1–15.

Lyons, O. Sacred Land Film Project. (2016a, September 16). *Indigenous view of the world* [Video file]. Retrieved from https://www.youtube.com/watch?v=kbwSwUMNyPU&list=PLgWo3ZPZn_6637eJnbskwnUyuA9I4Ztm-&index=13&t=1757s

Lyons, O. Sacred Land Film Project. (2016b, November 9). *Rights and responsibilities* [Video file]. Retrieved from https://www.youtube.com/watch?v=USwPW29W-aY

McLeod, N. (2007). *Cree narrative memory: From treaties to contemporary times*. Saskatoon: Purich.

Meyer, M. A. (2003). Hawaiian hermeneutics and the triangulation of meaning: Gross, subtle, causal. *Social Justice, 30*(4), 54–63.

Meyer, M. A. (2013). Holographic epistemology: Native common sense. *China Media Research, 9*(2), 94–101.

Mohawk, J. (1990, April 18). *Perspective of mother earth*. From WinterCamp chronicles. GENERATIONS native American Radio Archives. Boulder, CO. Retrieved from http://twoelk2.tripod.com/Generations/WCCJohnMohawk.html

Peschak, T. (2017, Aug 25). In this lagoon, the whales come to you. National Geographic. Retrieved from https://www.nationalgeographic.com/photography/proof/2017/08/gray-whales-baja-mexico

Porter, T. (2008). *And grandma said...Iroquois teachings as passed down through the oral tradition*. Bloomington, IN: Xlibris.

Smith, G. H. (2000). Maori education: Revolution and transformative action. *Canadian Journal of Native Education*, *24*(1), 57-72.

Smith, L. T. (1999). *Decolonizing methodologies: Research and indigenous peoples*. London: Zed.

Smith, L. T. (2000). Kaupapa Maori research. In M. Battiste (Ed.), *Reclaiming indigenous voice and vision* (pp. 225–247). Vancouver, BC: UBC Press.

Stewart-Harawira, M. (2005). *The new imperial order: Indigenous responses to globalization*. London: Zed.

Trudell, J. (2001). *They're mining us. In DNA (descendant now ancestor)*. London: Effective Records.

Vizenor, G. R. (1999). *Manifest manners: Narratives on postIndian survivance*. Lincoln, NE: University of Nebraska Press.

Wall, S., & Arden, H. (1990). *Wisdomkeepers: Meetings with Native American spiritual elders*. Hillsboro, OR: Beyond Words Publishing.

Wolfe, P. (2016). *Traces of history: Elementary structures of race*. London: Verso.

Chapter 9
The Hell of Lies, Denial, and Distraction: Critical Environmental Pedagogy Through Popular Dystopic Films

William M. Reynolds

> *[Trump] then tweeted: "It's really cold outside, they are calling it a major freeze, weeks ahead of normal. Man, we could use a big fat dose of global warming!" (Dreyfuss, 2015).*
> *Give me clean, beautiful and healthy air—not the same old climate change (global warming) bullshit! I am tired of hearing this nonsense (Trump, 2014).*
>
> *All calls for revolt, for halting the march toward economic, political, and environmental catastrophe, are ignored or ridiculed. Even with the flashing red lights before us, even huge swaths of the country living in Depression-like conditions, we bow slavishly before the enticing illusion provided to us by our masters of limitless power, wealth, and technological prowess. The system, although it is killing us is our religion (Hedges, 2015, p. 32).*

Introduction

This chapter discusses the importance of anthropogenic global warming and its representation(s) in Hollywood films. The environmental crisis is the major issue of our time and in many cases, the issues are simply denied by those who are opposed to science. The issues are kept from critical discussions in schools by this denial and the consequent controversy critical discussions of the concepts might elicit within the context of the curriculum of state mandated testing and the memorization of discrete bits of information quickly forgotten on the path to memorizing more discrete bits of information.

W. M. Reynolds (✉)
Georgia Southern University, Statesboro, GA, USA
e-mail: wrey@georgiasouthern.edu

A. S. Gkiolmas, C. D. Skordoulis (eds.), *Towards Critical Environmental Education*, Critical Studies of Education 14,
https://doi.org/10.1007/978-3-030-50609-4_9

> In other words, political conservatives have placed themselves in direct conflict with modern scientific knowledge, which shows beyond serious question that *global warming is real and caused by humans, and evolution is real and the cause of humans*. If you don't accept either claim, you cannot possibly understand the world or our place in it (Mooney, 2012, p. 7).

For those who are concerned and believe in science and in global warming, it is the existential crisis of our times, but in many ways, it is easy to forget, easy from which to be distracted or easy to shake our heads and wonder what we can possibly do about it.

The treatment of climate change and global warming in film is presented in various ways in various films. Most of the discussion on films concerning global warming and the environmental crisis centers on the documentaries that have been produced [*An Inconvenient Truth* (2006), *The Great Warming* (2016) and *Before the Flood* (2016)]. These documentaries magnify the global importance and ramifications of global warming and climate change, but receive only limited attention. In the case of Gore's documentary, there was intense initial interest, but it quickly faded. In the case of global warming and its post-apocalyptic Hollywood versions, three of the films analyzed are fictional representations. Fictional films on climate change tend to be dystopic views of the consequences or results of climate change or environmental destruction. One of the most well known of the films that portrays the dire consequences is *The Day after Tomorrow* (2004) with the protagonist paleoclimatologist, Jack Hall. The discussion of having a major motion picture hero being positioned as a paleoclimatologist is analyzed in the chapter. The other films chosen because of the dystopic orientation are *10 Last on Earth* (2019) and *Hell* (2012). These films and terminology will be analyzed to demonstrate that fictional Hollywood dystopic films have a type of redemptive quality within. One of the messages in these films is there will be a savior of some sort that will save the planet at the end or there will be a way out or a method of escape—Roughly translated: Don't worry everything will be fine. There is no need to worry. This is another blockbuster film characteristic.

As with other issues in critical media literacy and its connections with films and popular culture, this use of film could be brought into the school curriculum and critically discussed. Rebecca A. Martusewicz, Jeff Edmundson, and John Lupinacci in their insightful and provocative book, *EcoJustice Education: Toward Diverse, Democratic, and Sustainable Communities* (second edition) (2014), discuss issues of bringing concepts of ecojustice and climate change into the schools. The authors define ecojustice as follows:

> The understanding that local and global ecosystems are essential to all life; challenging the deep cultural assumptions underlying modern thinking that undermine those systems; and the recognition of the need to restore the cultural and environmental commons (Martusewicz, Edmundson & Lupinacci, 2014, p. 362).

The chapter concludes with the discussion of dystopic films centering on environmental disaster as it is connected to critical pedagogy.

The Status of "Climate Change" in the Era of Lies, Denial, and Distraction

> To tell the truth, to arrive together at the truth, is a communist and revolutionary act (Gramsci & Togliatti, 1919, p. 1).

Before analyzing Hollywood's visions and narratives of climate change, it is important to take a brief look at the reality of climate change in the present historical moment. Locally, the issues of climate change and the environment were heightened for me in 2018 when a development company decided to build a multiple dwelling complex on 19.2 acres of agricultural land in my county. Of course, we can't stop development, but this particular development was to be placed in proximity to a county Greenway which is a walking, running, and biking trail. The original purpose of which was to provide a natural/non-traffic environment for healthy exercise and to preserve the natural environment including a large variety of wildlife. The process involved several meetings with the Statesboro, Georgia City Council. A group—Friends of the Greenway—presented arguments in support of not developing the area. In the end, the developers won and the Greenway will not be as natural nor as safe. After winning the debate, the company decided to put the property up for sale and not construct the proposed development. In our neoliberal times, money and profit are much more important that preservation of the environment (more about this later in the chapter). There is also commercial development occurring on the Greenway as well. This consists of a warehouse looking Cross Fit Building. In order to prepare the ground for this, trees were clear cut and the land bulldozed. It reinforced the concepts for me that the environment is fragile and up to the highest bidder. Klein (2014) was correct. It is a question of capitalism versus the climate (environment). The issue for the Greenway in Bulloch County was not really an issue of environmental consciousness on the part of the developers or the City Council or even the chance of families obtaining affordable housing but about making money in this neoliberal moment. Destruction including the clear-cutting of trees to what purpose? This infuriating local example, of course, is only part of a much larger global issue. Yet, local issues have a way of making us more aware and attuned to larger complex issues. The era of Lies and Denial does not stop with a local issue, but is expanded nationally and internationally.

The US government released on November 23, 2018 *The Fourth National Climate Assessment Report*. It discusses the impact of climate change by region. It has a dire assessment. It is a frightening scenario. There are some significant predictions:

1. Crop production will decline. This will be due to higher temperatures, drought, and flooding.
2. Cattle will suffer.
3. Food sources from the sea will decline—ocean acidity and red tides.
4. Food and waterborne illness will spread. History will be lost.

5. Insect population will increase and diseases spread by insects will increase (Lyme disease, West Nile virus).
6. Asthma and allergies will be worse. Increase in CO_2 levels and increase in mold growth.
7. Serious mental health consequences—due to stress, anxiety, and depression.
8. Heat will cause an increase in deaths.
9. Work hours will decrease. The Southeast alone will probably lose 570 million labor hours by 2090.
10. Infrastructure (bridges, roads) will be negatively impacted by rising temperatures.
11. Water infrastructure will be challenged.
12. Floods will be more frequent.
13. Wildfires will increase.
14. History will be lost. The Southeast would lose 13,000 recorded historic and prehistoric archeological sites due to projected sea-level rise.
15. There will be more snakes and other invaders (Christensen, 2018, p. 1).

One can just imagine all the Hollywood blockbuster dystopic climate movies that could emerge from these issues and all the money that could be made from those films. Despite the seriousness of these horrible forecasts, President Trump had a reaction to the report:

> REPORTER: Have you read the climate report yet?
>
> TRUMP: I've seen it. I've read some of it and—it's fine.
>
> REPORTER: They say the economic impact could be devastating.
>
> TRUMP: Yeah. I don't believe it.
>
> REPORTER: You don't believe it?
>
> TRUMP: No, no. I don't believe it. And here's the other thing: You're going to have to have China and Japan and all of Asia and all of these other countries—you know, it addresses our country. Right now, we're at the cleanest we've ever been, and that's very important to me. But if we're clean but every other place on Earth is dirty, that's not so good (Bump, 2018, p. 1).

Climate Change Deniers

> With all of the hysteria, all of the fear, all of the phony science, could it be that man-made global warming is the greatest hoax ever perpetuated on the American people? It sure sounds like it (Inhofe, 2012, p. 1).

> I think there's assumptions made that because the climate is warming, that that necessarily is a bad thing. Do we really know what the ideal surface temperature should be in the year 2100? In the year 2018? I mean it's fairly arrogant for us to think that we know exactly what it should be in 2100 (Pruitt in Holden, 2019, p. 1).

> A former coal lobbyist, Donald Trump has nominated to run the Environmental Protection Agency on Wednesday, touted rolling back pollution standards and declined to identify climate change as a crisis requiring unprecedented action from the United States (Holden, 2019, p.1).

Although there has been a plethora of discussions, presentations, and publications about Fake News and post-truth, I prefer to use the term lies. The denial of the climate crisis is simply a lie, "a false statement made with deliberate intent to deceive; an intentional untruth; a falsehood" (see dictionary.com). Climate change deniers are aware that if they admit there is, indeed, a climate crisis then the free market (neoliberalism, capitalism) loses its justification. It requires a denial of science.

Trump's lies are obvious. But the capitalist/free market agenda and its connections to the climate crisis is a bit more complex than simply a politician making money or telling thousands of lies. The free market system and its advocates are afraid that the unfettered climate destruction that they have wrought on this earth will be unveiled.

> More fundamentally than any of this, though, is their deep fear that if the free market system really has set in motion physical and chemical processes that, if allowed to continue unchecked, threaten large parts of humanity at an existential level, then their entire crusade to morally redeem capitalism has been for naught. With stakes like these, clearly greed is not so very good after all. And that this is behind the abrupt rise in climate change denial among hardcore conservatives: they have come to understand that as soon as they admit that climate change is real, they will lose the central ideological battle of our time (Klein, 2014, p. 40).[1]

It is certainly about the environment versus the free market system and the money that fossil fuel companies and alike make from destroying the environment. It is more than that greed and avarice and the war against science and truth. Analyzing the major climate deniers and their political connections to one another is enlightening. The Republican politician most associated with climate change denial is James Inhofe, senator from Oklahoma. His famous senate speech on the hoax of climate change was followed in 2014 by Inhofe holding a snowball as he stood on the senate floor claiming that it was cold outside, which apparently was proof that climate change was a hoax. Scott Pruitt is a friend of Inhofe; both of them are from Oklahoma. Pruitt was an independent lawyer. His law service was entitled, Christian Legal Services. He had served as an Oklahoma State Senator from 1998 to 2000 and Oklahoma Attorney General from 2011 to 2017. His political positions in Oklahoma were against environmental regulations and a friend of the fossil fuel industry from which he received over $200,000.00 in contributions. As attorney general in Oklahoma, Pruitt sued the EPA more than 14 times. Of course, this was just the correct background for Trump to nominate Pruitt to administer the Federal Environmental Protection Agency. After continuing his orientation for deregulating environmental protections and climate change denial, he was overcome with sandal and misdeeds; he resigned the office in July 2018. Quickly, Trump nominated his replacement, Andrew Wheeler, a coal industry lobbyist, in July of 2018 as Acting Administrator of the EPA. He also has a connection to James Inhofe and is a climate change denier.

---

[1] See also Klein, N. On fire: The (burning) case for a green new deal. New York: Simon & Schuster.

> Wheeler, a principal at Faegre Baker Daniels Consulting, is a lobbyist for coal giant Murray Energy and served as a top aide to Sen. James M. Inhofe (R-Okla.) when Inhofe chaired the Senate Environment Committee. He has represented Murray Energy—whose chief executive, Bob Murray, is a prominent supporter of the president—since 2009 (Eilperin & Dennis, 2017).

This is just a small foray into the political machinations of climate deniers in the political spectrum. In "2017 53 out of 100 members of the US Senate were climate change deniers along with 232 of 435 members of the House of Representatives"(Emerson, 2017, p. 1). Emerson clarified what she meant by climate deniers:

> For the purpose of this survey, we defined climate change deniers as those who deny the existence of anthropogenic, or human-made, climate change. Senators and representatives who called themselves "skeptics" were also included, because enough empirical evidence exists for them to make an informed decision on whether people are influencing the climate. Both groups include Republicans and Democrats, though GOP members largely outnumber their counterparts (Emerson, 2017, p. 1).

These free market fundamentalists are also connected to fundamentalists of another kind. Although given the current administration's activities, the connection(s) are a bit more twisted. The free market, greed, and profit are the underlying connection.

There also exists an association among climate deniers and right-wing Christianity, the Civic Gospel and Christocrats (Reynolds & Webber, 2009).[2] The connection exists with Biblical interpretations particularly on the Book of Revelation.

> Many Christocrats feel that concern for the future of our planet is irrelevant, because it has no future. They believe we are living in the End Time, when the son of God will return, the righteous will enter heaven, and sinners will be condemned to eternal hellfire. They may also believe, along with millions of other Christian fundamentalists, that environmental destruction is not only to be disregarded but actually welcomed—even hastened—as a sign of the coming Apocalypse (Scherer, 2004, p.11 in Reynolds & Webber, 2009, p. 151).

In *Divine Destruction: Wise Use Dominion Theology, and the Making of American Environmental Policy* (2005), Hendricks discusses End Times and the rejection of environmental concerns. Through her study, she found that some Christocrats believe that causing environmental destruction is a way to hasten the End Times and the return of Christ. The notion is based on the Biblical passage in Genesis that states that man should have dominion over the earth. This is part of dominion theology (see Reynolds & Webber, 2009). Hendricks indicates how this view contributes to the disregard for the environment.

> This view, when combined with the belief that the End Times are near, leads some to believe that either there is no need to take care of the environment, or, alternatively, that actively exhausting the environment will speed the Second Coming (Hendricks, 2005, p. 49).

---

[2] Christocrats is a term that refers to those conservative right-wing Christians who combine religion and politics to move toward a Christian Nation. Ironically, it melds with and supports Trump's Make America Great Again.

There is climate change denial in the support of the free market/capitalism and there is a type of connected disregard for the environment with Christocrats. These denials (lies) are reinforced by media both with the visibility of deniers and climate denial films.

There is climate denial in films. In the case of denial films, most are "documentaries." One film can serve as an example. As we examine the impact of media and its critical analysis, a look at a representative documentary of climate deniers is important. *The Great Global Warming Swindle* (2007) is perhaps the most well-known denier documentary. This film declares that human-made global warming is the "biggest swindle in modern history." The film indicates ironically that the entire notion of global warming is based upon funding and politics.

The Great Global Warming Swindle

The Great Global Warming Swindle (2007) is a problematic documentary that premiered on March 7, 2007 on the United Kingdom's Channel 4. It proclaimed that the entire scientific community and its opinion on global warming are not based on science but come from the influences of outside funding and political commitments. It raises issues centered on whether there actually is scientific agreement concerning anthropogenic global warming. The film's original title was *Apocalypse my arse.* The film was directed by David Durkin, a former Marxist.

> In his program, Mr. Durkin rejects the concept of man-made climate change, calling it "a lie ... the biggest scam of modern times." The truth, he says, is that global warming "is a multibillion-dollar worldwide industry, created by fanatically anti-industrial environmentalists, supported by scientists peddling scare stories to chase funding, and propped up by compliant politicians and the media." (Webb, 2007, p.1).

In the film, there are a number of "experts" from various fields from environmentalists to academics who support the notions that climate change is a lie and various other assertions made in the film.[3]

They are interviewed to demonstrate and give credibility to the claims of global warming being a type of for-profit scam assisting Western environmentalists. Of course, the film stirred up controversy. But, perhaps the best way to summarize this film is that *The Great Global Warming Swindle* is itself a swindle. In an article written by George Monboit in *The Guardian*, the author summarizes the entire film and subsequent controversy succinctly:

---

[3] Some of the experts interviewed in the film: Fred Singer, emeritus professor of environmental science at the University of Virginia; Roy Spencer, research scientist at the University of Alabama in Huntsville; John Christy, a climate scientist at the University of Alabama in Huntsville; Patrick Moore, former Canadian Greenpeace President; Patrick Michaels, a climatologist; Nir Joseph Shaviv; an Israeli/American physics professor and climate change denier; Richard Lindzen, an American atmospheric physicist. Many other scientists, weather forecasters, meteorologists, and academics are interviewed in the film.

> Cherry-pick your results, choose work which is already discredited, and anything and everything becomes true. The twin towers were brought down by controlled explosions; MMR injections cause autism; homeopathy works; black people are less intelligent than white people; species came about through intelligent design. You can find lines of evidence which appear to support all these contentions, and, in most cases, professors who will speak up in their favour. But this does not mean that any of them are correct. You can sustain a belief in these propositions only by ignoring the overwhelming body of contradictory data. To form a balanced, scientific view, you have to consider all the evidence, on both sides of the question (Monboit, 2007, p. 1).

This film is part of the effort to delegitimize anthropogenic global warming. It is frequently referenced by deniers. But its credibility is clearly suspect. From documentary global warming deniers, we turn to the influence of popular films on perceptions of this environmental crisis. Hollywood attempts in many ways to minimize the crisis.

Popular Films: The Infection of Cultural Hegemony, the Manufacture of Consent, and Critical Media Literacy

> The "spontaneous" consent given by the great masses of the population to the general direction imposed on social life by the dominant fundamental group; this consent is "historically" caused by the prestige (and consequent confidence) which the dominant group enjoys because of its position and function in the world of production (Gramsci, 1971, p 12).
>
> The notion of hegemony, as I have suggested before, refers to a form of ideological control in which dominant beliefs, values, and social practices are produced and distributed throughout a whole range of institutions such as schools, the family, mass media, and trade unions. As the dominant ideology, hegemony functions to define the meaning and limits of common-sense as well as the form and content of discourse in a society (Giroux, 1981, p. 94).
>
> Hegemony, for Gramsci, refers to how ruling elites, through the organs of mass culture, manipulate our understanding of reality to promote their interests. The passive consumers of mass culture see the world not as it is but as it is interpreted for them. Mass culture, including the press, schools and systems of entertainment, demonizes all those the ruling elites scapegoat and fear—in our case people of color, the poor, Muslims, undocumented workers, anti-capitalists, labor unions, intellectuals, liberals and dissidents (Hedges, p. 1).

Popular Hollywood films on climate change are part of the infection of the mass culture and hence contribute or are part of the elite cultural hegemony and consequently the distraction from a critical and genuine concern and pedagogy for the planet and result in an unrecognized voluntary compliance, manufactured consent, or spontaneous consent. Media, in this case Hollywood films, demonstrate the power of cultural hegemony. Citizens become so preoccupied with making a living, and consuming the latest technological products that Hollywood films offer a path of distraction and consent to the dominant hegemony which is not only cultural, political, and ethical but also economic. The cultural hegemony in terms of global warming, climate change, or planetary disaster is don't worry, be happy because either there will be an escape or this was meant to be and heaven is just an apocalypse away.

In the Hollywood films I am analyzing, there seems to be at least three common characteristics.

First, there is always a smart enough technician or scientist that can and will rescue and save us. These scientists/technicians tend to be white males (the notion of white males being the savior is the subject for additional research). There are exceptions of course. Micah in *IO* is an African American male and Sam the main character is a woman. There are also women scientists in films like *Contact* (1997) with Jody Foster. Second, no matter what the climate disaster, there is always a cure or escape—shuttles, transportation to safe places, and so on. Third, human kindness alone can overcome global climatic apocalypse. The following films all demonstrate these characteristics.

The Day after Tomorrow (2004)

One of the most famous films concerning climate disaster is *The Day After Tomorrow* (2004). It is based on the book, *The Coming Global Superstorm* (1999) by Art Bell and Whitley Strieber. The film depicts catastrophe climate effects following the disruption of the North Atlantic Ocean circulation in a series of extreme weather events that usher in global cooling and lead to a new ice age (Gillis, 2016).

The film begins with Jack Hall an American paleoclimatologist and his two fellow scientists drilling for ice-core samples in the Larsen Ice Shelf for the National Oceanic and Atmospheric Administration (NOAA) as they are working on the project the shelf collapses. After almost losing his life, Jack ventures to the United Nations Global Warming Conference in New Delhi where it is snowing! Jack warns that there will be a climate shift although he is uncertain about when it will occur. After his address, the Vice President of the United States, Raymond Becker, rejects and ridicules Jack's apprehensions. During the conference Terry Rapson of the Hedland Centre in Scotland teams up with Jack based on their shared views of climate change.

The truth wins out as gigantic storms envelop the world particularly the Northern hemisphere, the United States, Canada, Siberia, and Scotland. The temperature drops to −150 °F. The world is caught in massive climatic disasters like gigantic hailstorms, rising sea levels, tornados, and so on. President Blake grounds air traffic and orders the populations of the southern states to be moved to Mexico. The storms in Europe also claim the lives of Jack's colleague Rapson and his team.

Meanwhile Jack's son, Sam, and his friends, Brian Parks and Laura Chapman, are in New York City for an academic contest. They experience a ferocious storm that floods Manhattan. Eventually, the flood waters freeze. Jack has told Sam to stay inside the public library where he, his friends, and many others have taken shelter. When the water freezes many of the survivors in the library decide to hike south. They are later found dead from hypothermia. They are found by Jack and his colleague Jason as they make their way to Manhattan, which is frozen solid. The others have stayed in the library and burn books to keep warm in the frigid weather. Later the storm(s) abate and Sam's group is found alive and rescued by Jack and Jason, who have had a precarious journey to New York City. One of Jack's friends perishes to save Jack and Jason during the trip.

In the end, Jack's son is rescued from Manhattan along with many other survivors who await helicopter transport. The President now Raymond Becker, since President Blake is killed when his motorcade is caught in the storm, apologizes for his mistake of denying climate change; he issues orders for helicopters to pick up survivors in the northern areas of the United States. Finally, the astronauts in the space station view a new Earth with ice caps across the Northern Hemisphere.

Again, all is well that ends well. Even with a devastating new ice age and countless causalities, everything will be alright. The politicians including the President will see the error of their ways, apologize, and send rescue helicopters to save survivors. The kids will survive a catastrophe freeze in New City and be saved by a scientist father and his colleagues. And the world will go on and human kindness will overcome the devastation of climate change.

A bit less romantic ending occurs in the film *Hell*. But, hope endures nonetheless. This time it is a scorched Earth rather than a frozen one.

Hell (2011)

> The sun will have scorched everything (Emmerich, Walther, Wobke, & Fehlbaum, 2011).

Hell (2011) is a German-Swiss film. It is in the post-apocalyptic, horror, science fiction genre. The executive producer of the film, Roland Emmerich, is associated with films such as *Independence Day* (1996) and *2012* (2009). This film is the darkest of the three films discussed. It takes place in a post-apocalyptic Germany after the Earth's atmosphere has been destroyed by solar flares. The Earth's temperature has risen by 50 °F. The Earth has been devastated. Food and water are in short supply. Civilization has basically collapsed.

The film follows the journey of three individuals: Marie a young woman, her sister Leonie and Phillip, a love interest of Marie. The three are driving in an old Volvo constructed to protect them from the sun. Only a small slot in the windshield of the car is allowed to provide driver visibility. They are trying to make it to the mountains and the water is said to be there. Their trip continues when they stop at a gas station to find gas, food, and water. Tom arrives on the scene and attempts to steal food and water from the Volvo. Philip and Tom get into a fight. The fight ends, and Tom and the trio agree that they will bring Tom along because he is a mechanic and can fix the car. During their journey, they are captured and through a number of violent fights with a group of farmers, who have become cannibals, the two sisters manage to survive. Phillip, on the other hand, is executed in the "slaughterhouse."

The three survivors, Marie, Leonie, and Tom make it to the mountains. There they find water. They see birds in the sky even though the rest of the landscape is ravaged and blighted. The film leaves us with a sense that, even in this horrible apocalypse, there is hope that life will survive and it will all get better. This film differs from the *Day after Tomorrow* and *IO*, in that there is no superhero scientist that will solve the decimation of the Earth nor are there rocket ships to safety. This film is bleaker than that. Yet, at the end of the film, there is a glimmer of hope. Everything, it seems, will be ok in the future.

IO Last on Earth (2019)

The last film to be discussed is *IO* in the post-apocalyptic science fiction genre. It debuted on Netflix in 2019. It is an interesting film and different than many films in the genre but contains many of the same messages.

At the beginning of this film, most humans have left the planet on Exodus shuttles to a space station near Jupiter's moon IO to survive the toxic devastation on Earth. The exact cause of the toxic atmosphere is never specifically explained, but the viewer understands that the air on Earth is no longer breathable except in limited high elevation areas. Otherwise, survival requires survivors to wear masks and oxygen tanks. Sam, the daughter of Dr. Harry Walden (his area of expertise is never stated), lives alone and tries to raise bees that can survive the toxic atmosphere and pollinate plants that can create oxygen. Her father had advocated that humans stay on Earth because he still believed there was hope. This is the hope that helps her to survive. She manages to grow vegetables as her source of sustenance.

One day a man, Micah arrives in helium balloon on his way to get to the launch site of the final Exodus shuttle leaving Earth. He arrives after a storm has gone through Sam's beehives and basically decimated all of the bees. Micah has actually come to find Dr. Walden to "speak" with him about Walden's advocacy of staying on Earth. Micah's wife died because of staying on Earth and not having enough to eat. A short bit of dialogue from the film indicates the feelings of the characters.

> Micah: We weren't ready for this.
> No matter we got from people like your father.
> What we saw with our own eyes,
> The future got unthinkable for everyone.
> Sam Walden: We thought we could find a way for humans to survive on Earth.
> We were wrong.

Sam and Micah become involved romantically, a Hollywood device to help assure viewers that everything is normal and things will be fine. Eventually, both a reluctant Sam and Micah decide to leave on the final shuttle and must go to the toxic (unnamed) city to acquire more helium for Micah's balloon since the shuttle launch site was moved and is a longer distance away. Sam leaves Micah and travels to the Art museum where she pauses at a painting, *Leda and the Swan* which inspired a poem of the same title by Yeats. The painting and the story center on the god, Zeus disguised as swan, impregnating Leda. She becomes the mother of Helen of Troy. The connection of the pregnancy is a bit unclear at this point in the film. Sam decides to take off her mask and breathe the toxic air. The ending of the film is Sam standing on a beach in a breeze with a young black child, apparently Micah's. So, it ends on a hopeful note that there is still hope for people to return to Earth.

So, the film ends with a comforting note that, despite the fact that Earth's atmosphere might be devastated by mysterious toxic waste and thousands died as a result of this environmental disaster, not to worry there will be scientists and technologists who can get us off the planet and there might be a miracle and eventually we can return to a new clean earth.

These three films demonstrate that serious critical consideration of media needs to be pursued. This critical analysis would assist viewers of films to understand the hegemonic messages contained. It should be part of a critical pedagogical education which questions the neoliberal logic of free market fundamentalism in the present moment.

Global Warming and Critical Pedagogy Through Critical Media Literacy

> The revolutionary leaders of every epoch who have affirmed that the oppressed must accept the struggle for their liberation—an obvious point—have also thereby implicitly recognized the pedagogical aspect of this struggle. Many of these leaders, however (perhaps due to the natural and understandable biases against pedagogy) have ended up using the 'educational' methods employed by the oppressor. They deny pedagogical action in the liberation process, but they use propaganda to convince (Freire, 2006, pp. 67–68).
>
> The greatest existential crisis of our time is to at once accept the tragic reality before us and find courage to resist. It is to acknowledge that the world as we know it will become harsher and more difficult, that human suffering will expand, but that we can, if we fight back, perhaps reconfigure our lives and our society to mitigate the worse savagery, dramatically reduce our carbon footprint and save ourselves from complete annihilation. The power elites will do nothing to save us (Hedges, 2017, p.3).

Present-day schooling is locked into accountability, standards, high stakes testing, Pearson packaged curriculum, textbooks, Craigslist graders, and on and on. An education (not schooling) based on criticality is dismissed, denied, and actively opposed. Schooling has become a disconnected oppressive exercise that for the most part ignores the socio-economic-political aspects of society in favor of an almost exclusive focus on getting a job. Securing employment is important, but at what cost? This type of schooling takes place in elementary and secondary schools and insidiously has worked its way into teacher education as well as other areas in universities. At this historical moment, schools are locked into corporate-produced and -graded high-stakes exams, and colleges of teacher education are hopelessly locked into "training" preservice teachers to teach to the tests. This is touted as the way to succeed in the profession. Not only has this occurred in teacher education at universities but also critical education continues to disappear from the scope of schooling at universities in general.

> A disconcerting number of academics and teachers in the current moment continue to join forces with right-wing politicians and conservative government agencies to argue that class-

> rooms should be free of politics. Their shared conclusion? That schools should be spaces where matters of power, values and social justice should not be addressed (Giroux, 2019, p. 1).

The elimination of philosophy, theory, and social foundations from university course offerings is an example. Many teachers and professors voluntarily comply with the demise of any type of critical education or any attempt to raise critical consciousness. Power operating is not simply a matter of "coercion or repression, the domination of one group of people by another… power requires a degree of complicity on the part of the ruled to function" (Buchanan, 2008, pp. 13–14). This acquiescence is the complier's effort toward job security and advancement. This is a fundamental political problem or question of our time. Deleuze considered the intricacies of voluntary subservience to be a central political problem, "which was to preoccupy both Deleuze and Guattari for most of the rest of their lives" (Buchanan, 2008, p. 7). This voluntary subservience has increased in the twenty-first century. As Giroux indicates, it has resulted in the movement toward a "politically free" (Giroux, 2019) education.

In terms of global warming, educators need to fight back against the exclusion of ecojustice and other issues of social justice in the curriculum and the subsequent lack of dialogical community in general. There is confusion and misunderstanding about community (a necessity when thinking about ecojustice) and social networks. Zygmunt Bauman describes the difference in this way:

> The question of identity has changed from being something you are born with to a task: you have to create your own community. But communities aren't created, you have one or you don't. What social networks can create is a substitute. The difference between a community and a network is that you belong to a community, but a network belongs to you. You feel in control (Bauman in De Querol, 2016).

Any critical pedagogy, at this moment of the bleak future offered by global warming/climate change, must not only involve a critical media literacy analysis of films, and other media centering on global warming, but also incorporate local involvement and struggle with issues of the blatant disregard of the environment for the sake of profit. The capitalist/neoliberal hunger for profit and money must be resisted by those interested in preserving the earth. This can be done at the local level with critical pedagogues encouraging and assisting students in engaging in involvement with environmental issues and larger global issues as well (see Martusewicz, Edmundson, & Lupinacci, 2011). I found struggling for environmental issues at a local level contributes to a heightened awareness of the necessity of critical analyzing and struggling with those environmental issues on a global level. It is, as Klein suggests, a struggle between capitalism and the climate. The tragic truth in the time of lies is that the elimination of critical education makes the destruction of the planet more likely, increased consuming and ignoring social justice more likely. This lack and its perpetuation also assure the continuation of the education of a type of consumer citizen and not an engaged critical citizen. We are witnessing the result of that type of education in our everyday lives at this historical moment.

Cultural Hegemony Makes Us Feel Good in a Bleak Future of Global Warming

> In other words, the culture that triumphed in our corporate age pits us against the natural world. This could easily be a cause only for despair. But if there is a reason for social movements to exist, it is not to accept dominant values as fixed and unchangeable but to offer other ways to live—to wage, and win a battle of cultural world views (Klein, 2014, pp. 60–61).

Two popular songs demonstrate in many ways the hegemonic take on anthropogenic global warming—Don't Worry Be Happy (1988) by Bobby McFerrin and Happy (2015) by Pharrell Williams. All of the cultural and political icons of the neoliberal agenda are telling us not to worry about the environment. The first reaction is to simply deny the existential crisis. Secondly, focus on buying things, in this particular case, products that are environmentally disastrous. The focus is to ignore the environment in the quest for more cell phones, big trucks, or SUVs with low gas mileage, plastic water bottles, round-up weed killer, and so on. These distractions of consuming help to construct the current hegemony. As we have seen, even Hollywood movies dealing with catastrophic global warming leave us with a sense that this whole global warming issue is something that will be handled or there will be an escape. In one sense, the entire discussion and representation of global warming can be made into a for-profit phenomenon without any critical analysis. Feeling troubled about global warming? Not a problem. Go to the movies that will make you feel better about it. Put up a Facebook post about it, contribute to a type of slacktivism. Buy a t-shirt expressing some aspect of it. There must be an alternative. Part of that alternative is a public pedagogical responsibility. I agree with Naomi Klein when she writes.

> Fundamentally the task is to articulate not just another alternative set of policy proposals but an alternative worldview to rival the one at the heart of the ecological crisis—embedded in interdependence rather than hyper-individualism, reciprocity rather than dominance, and cooperation rather than hierarchy (Klein, 2014, p. 462).

Again, these changes to an alternative worldview cannot possibly be achieved with an uncritical look at our culture, with a "politically free" education, or with a don't-worry-be-happy vision of the future. Critical pedagogy coupled with critical media literacy is one way to begin the journey to a new view, a view that confronts the deniers and changes test-driven education. I would rather believe that there is a modicum of hope. It is difficult to have hope in this time of lies, but it is essential.

References

Bell, A., & Strieber, W. (1999). *The coming global superstorm*. New York, NY: Atria Publishing Group.

Berman, M. B., Rister, L., & Helpert, J. (2019). *IO [motion picture]*. Los Angeles, CA: Mandalay Pictures/ Netflix.

Buchanan, I. (2008). Power, theory and practice. In I. Buchanan & N. Thoburn (Eds.), *Deleuze and politics* (pp. 13–34). Edinburg: Edinburg University Press.
Bump, P. (2018, November 26). *Trump responds to his administration's report indicating a huge cost from climate change: 'I don't believe it'*. Retrieved from https://www.washingtonpost.com/politics/2018/11/26/trump-esponds-report-indicating-huge-cost-climate-change-i-dont-believe-it/?noredirect=on&utm_term=.69eb1470e3cb
Christensen, J. (2018, November 26). *15 takeaways from the US climate change report.* Retrieved from https://www.cnn.com/2018/11/26/health/climate-change-report-15-takeaways/index.html
De Querol, R. (2016). *Interview Zygmunt Bauman: "Social media are a trap"*. Madrid: El Pais. Retrieved from https://elpais.com/elpais/2016/01/19/inenglish/1453208692_424660.html
Dreyfuss, B. (2015, October 19). *Donald Trump still does not understand how seasons work.* Retrieved from motherjones.com: http://www.motherjones.com/blue-marble/2015/10/donald-trump-rhymes-with-ronald-grump
Durkin, M. (2007). *The great global warming swindle [motion picture]*. London: WAGtv.
Eilperin, J., & Dennis, B. (2017, July 21). Trump to tap longtime coal lobbyist for EPA's no. 2 spot. *The Washington Post.* Retrieved from *https://www.washingtonpost.com/news/energy-/*wp/2017/07/21/trump-to-tap-longtime-coal-lobbyist-for-epas-number-two-/?utm_term=.eb9dad933fd5
Emerson, S. (2017, April 28). The climate change deniers in congress. *Motherboard.* Retrieved from https://motherboard.vice.com/en_us/article/pg5zqg/a-guide-to-the-climate-change-deniers-in-congress
Emmerich, R., Gordon, M., & Emmerich, R. (2004). *The day after tomorrow* [motion picture]. Los Angeles, CA: 20th Century Fox.
Emmerich, R., Walther, G, Wobke, T., & Fehlbaum, T. (2011). *Hell* [motion picture]. Unterföhring: Paramount Pictures.
Freire, P. (2006). *Pedagogy of the oppressed: 30th anniversary edition.* New York, NY: Continuum.
Gillis, J. (2016, March 22). Scientists warn of perilous climate shift within decades, not centuries. *The New York Times.* Retrieved from https://www.nytimes.com/2016/03/23/science/global-warming-sea-level-carbon-dioxide-emissions.html
Giroux, H. A. (1981). *Ideology, culture and the process of schooling.* Philadelphia, PA: Temple University Press.
Giroux, H. A. (2019, May 13). If classrooms are "free of politics," the right wing will grow. *Truthout.* Retrieved from https://truthout.org/articles/if-classrooms-are-free-of-politics-the-right-wing-will-grow/
Gramsci, A. (1971). *Selections from the prison notebooks.* New York, NY: International Publishers.
Gramsci, A., & Togliatti, P. (1919, June 21). Worker's democracy. *L'Ordine Nuovo, 1*(7). Retrieved from https://www.marxists.org/archive/gramsci/1919/06/workers-democracy.htm
Hansler, J. (2018, February 12) *EPA administrator Scott Pruitt suggests climate change could benefit humans.* Retrieved from https://www.cnn.com/2018/02/08/politics/scott-pruitt-climate-change/index.html
Hedges, C. (2015). *Wages of rebellion: The moral imperative of revolt.* New York, NY: Nation Books.
Hedges, C. (2017). *Antonio Gramsci and the battle against fascism.* Portland, OR: Common Dreams. Retrieved from https://www.commondreams.org/views/2017/06/05/antonio-gramsci-and-battle-against-fascism
Hendricks, S. (2005). *Divine destruction: Wise use, dominion theology, and the making of American environmental policy.* Hoboken, NJ: Melville House Publishing.
Holden, E. (2019, January 6). Andrew Wheeler: Trump's EPA pick says climate change 'not the greatest crisis'. *The Guardian.* Retrieved from https://www.theguardian.com/environment/2019/jan/16/andrew-wheeler-climate-change-trump-epa-hearings
Inhofe, J. (2012). *The greatest hoax: How the global warming conspiracy threatens your future.* Washington, DC: WND Books.

Klein, N. (2014). *This changes everything: Capitalism and the climate*. New York, NY: Simon and Schuster.

Martusewicz, R. A., Edmundson, J., & Lupinacci, J. (2011). *EcoJustice education: Toward diverse, democratic, and sustainable communities*. New York, NY: Routledge.

Monboit, G. (2007, March 12). Don't let truth stand in the way of a red-hot debunking of climate change science. *The Guardian*. Retrieved from https://www.theguardian.com/commentisfree/2007/mar/13/science.media

Mooney, C. (2012). *The republican brain: The science of why they deny science – And reality*. Hoboken, NJ: John Wiley & Sons.

Reynolds, W., & Webber, J. A. (2009). *The civic gospel: A political cartography of Christianity*. Rotterdam: Sense Publishers.

Scherer, G. (2004). *The godly must be crazy: Christian-right views are swaying politicians and threatening the environment*. Retrieved December 12, 2007, from http://www.grist.org/news/-/10/27/scheerer-christian/

Trump, D. J. (2014). Tweet. 10:44 PM, January 28, 2014.

Webb, A. (2007, March 6). Global warming labeled a 'scam'. *The Washington Times*. Retrieved from https://www.washingtontimes.com/news/2007/mar/06/20070306-122226-6282r/

Chapter 10
Toward an Interspecies Critical Food Systems Education: Exploring an Intersectional Cultural Humility Approach

Teresa Lloro

Introduction

I shared my life with farm animals as a kid. Indeed, some of them, like my duck Sammy, my goat Fancy, and my sheep Cutey were among my best friends. My father, a sheepherder or *pastor* in Spanish, grew up in the 1940s in a tiny rural village in the Pyrenees mountains of northern Spain when the landscape was still mostly untouched by large-scale agriculture and mountain ski resorts. He immigrated to the United States in the 1960s and like many diaspora, carried with him centuries-old cultural traditions, including those involving animals and food. From a very young age my sister and I tended our large suburban vegetable garden with my father and also participated when my father slaughtered animals—mostly sheep and sometimes rabbits. We also spent time on my father's friends' dairies; there we would not only interact with fellow Spanish and Basque diaspora, but we also spent time with dairy cows—and the overwhelming number of houseflies that populate multispecies communities with concentrated excrement. I recall how the pungent smell would sting my eyes and nose when we first arrived, but after a bit of time there my olfactory receptors would adapt and the odor would seem to dissipate. My understanding of and sustained experiences with that entire corporeal and visceral process has significantly influenced my thinking about the use of animals for food, and in turn my pedagogy, especially because I regularly teach about animal agriculture in several different courses. While Wright-Maley perhaps correctly notes that "…becoming more closely connected to our food—raising one's own animals for instance—may lead us to rationalize and normalize our consumption of them as part of the normal order" (Wright-Maley, 2011, p. 4), for me, this close connection had the opposite effect: it inspired not only my vegetarianism and veganism, but also my

T. Lloro (✉)
California State Polytechnic University, Pomona, CA, USA
e-mail: tllor001@ucr.edu

A. S. Gkiolmas, C. D. Skordoulis (eds.), *Towards Critical Environmental Education*, Critical Studies of Education 14,
https://doi.org/10.1007/978-3-030-50609-4_10

animal advocacy and activism in many other aspects of my life (Lloro-Bidart, 2015; Lloro-Bidart & Semenko, 2017). Likewise, it has tremendously shaped my thinking related to cultural practices surrounding food animals, as well my role as an activist, educator, and researcher.

Growing up in suburban southern California in the 1980s and 1990s, our family traditions revolved around plants, animals, and food in ways that differed quite markedly from the norm. The middle and upper class White suburban and urban elite had not yet made keeping backyard chickens in a coop fashionable or associated with sustainable practice (Blecha & Leitner, 2014). Indeed, most of the families I knew who raised chickens were Mexican diaspora or Mexican-American—and they allowed their chickens to run free like we did. Women's ability to "break the glass ceiling" to become high-paid butchers was not yet a thing (Inge, 2014) and no one seemed to care about "meeting their meat" (Werft, 2015) or being a "conscientious omnivore" (Pollan, 2006; Rothberger, 2014) and "locavore" (Rudy, 2012). Most people loved my family's homemade lamb concoctions (lamb stew, barbequed lamb chops, and lamb meatballs) because they tasted good, not because they believed they were participating in humane farming (our sheep lived a really good life compared to most farm animals and they *were* loved) or because they were doing something great for the planet (our sheep's excrement fertilized our vibrant vegetable garden, they never had to travel long distances to slaughter and then to market, and we used nearly every part of the animal, including organ meats). So while it was certainly true that most of my friends and their families did not have farm animals in their backyards and bought all of their food at the local grocery store or market, we never viewed ourselves as being more sustainable than they were because of our access to freshly slaughtered, "humane meat." We just carried out the cultural traditions handed down to my father by his parents and grandparents, albeit in a different country and on a smaller scale.

Some of the fondest memories of my childhood involve sheep and lamb. I purposely use both of these terms because the former implies a live animal and the latter a young live animal or meat. I absolutely adored all of our sheep. Even though my father repeatedly warned me not to become too attached to them—especially the rams who could become quite aggressive as they aged—I stubbornly refused. Every so often my father's friend would pull up to the house with his old brown pick-up truck and I could not wait to glimpse the new wooly faces with twinkling eyes peeking through the bars of the truck, imagining all the fun we would have together. Since we did not have horses, but we knew our maternal grandfather used to race them, my sister and I would sometimes try to ride the large sheep like horses. Of course, this was not met with much success since the sheep usually resisted and balancing on their backs was quite difficult. I also remember hiding in my bedroom as a teen sobbing when one of my favorite sheep was slaughtered—a process that I no longer participated in by that time. And yet, I also still have fond childhood memories of hand-grinding fresh lamb meat with minced herbs and garlic from our garden to make my Dad's famous (and delicious) homemade lamb meatballs—*rebozada* style. I can still smell them now. I can also still smell the pungent odor of sheep wool, which leaves a greasy residue on your fingers whenever you stroke them. Lanolin, a wax that wool-bearing animals secrete through their sebaceous glands, is prized for all sorts of medicinal and topical uses, but I never wanted to smell like that when I went to school and had to interact with my friends who did not have farm animals. Yet, I also secretly cherished its stench because it always reminded me that I had the opportunity to share my life with other animals. Thus, my childhood memories and identity, even as an adult, are tightly bound not only sheep, but also lamb (as meat).

Clearly, the scenario I describe above does not reflect most contemporary forms of animal agriculture in the US, as many critics highlight (e.g., Adams, 1990/2015; Darst & Darst & Dawson, 2019; Joy, 2010; Potts, 2016). Concentrated animal feeding operations (CAFOs), which treat animals and low-income and/or racialized people like inanimate and nonliving parts of the industrial food chain (Food Empowerment Project, 2018; Pachirat, 2011), not only produce the majority of the meat in the US, but also contribute to significant environmental degradation and human suffering in surrounding communities (Ackerman, Musil, McAuliffe, Brunson, & Reynolds, 2017; Darst & Dawson, 2019; Harper, 2012; Kim, 2015; Potts, 2016; Russell & Semenko, 2016). Although small working family farms and sustenance farms can significantly reduce animal mistreatment through the provision of greater amounts of space, higher quality food, and sometimes physical care, animals ultimately die in any of these farms. Further, the meat and vegetables produced there are typically quite costly, making these products out of reach for low-income communities. And finally, small family farms can but do not necessarily provide better working conditions for the people they employ, though it is often assumed to be the case. Therefore, in sharing my own personal stories here, I certainly do not intend to romanticize what life and death were like for the sheep we raised or to imply that small family farms are the panacea for all of the perils of industrialized animal agriculture. Rather, I seek to highlight the multiple and overlapping contingencies and tensions associated with all forms of animal agriculture, even those that appear as "humane" (Pollan, 2002). I also want to explore what it means to engage with these contingencies and tensions in environmental education (EE).

As many others in education and beyond have poignantly noted (Lupinacci & Happel-Parkins, 2018; Rice, 2013; Rowe, 2013; Stapleton, 2015), and as the opening vignette of this article demonstrates, eating—including the act of eating the flesh of another animal— is deeply sensorial and visceral, as well as intimately tied to identity and sociocultural traditions. One does not need to have participated in animal slaughter the way I did as a child to know and understand that eating animal flesh *is* culturally significant. Countless people all over the US, for example, gather to celebrate birthdays, holidays, weddings, and other events while enjoying meat; for some, slaughtering a particular animal is a focal point of the event. While some people eat meat because it aligns with their taste preferences, because it is cheap and convenient, or simply because it is the norm in a carnist society (Joy, 2010), there is no denying that for many, it is incredibly culturally significant (e.g., Gaard, 2001; Kim, 2015; Stapleton, 2015). Thus, eating also implicates all human beings, regardless of class, culture, gender, race, or sexuality in complex relations with other humans, animals, and the environment, as everyone must eat and most people eat animals.

Critical animal-focused education (AFE) scholars in some ways push back against the importance of the cultural significance of eating animals by instead emphasizing power dynamics. That is, they point to the ways in which animals almost universally have less power than most humans, particularly food animals (e.g., Corman, 2017; Oakley et al., 2010; Pedersen & Stănescu, 2012; Rice, 2013; Russell, 2005; Russell & Fawcett, 2013; Spannring, 2017; Spannring & Grušovnik, 2018). Such tensions were well on display in the opening of this chapter as I described fond childhood memories of making and eating lamb meatballs, while also recounting the torment I experienced as a teen when my father killed my wooly friends, who had notably less power than we did. I share these childhood experiences to foreground the cracks and ruptures of, on the one hand theorizing eating animals as a sociocultural practice imbued with history, tradition, and identity, as illustrated in my description of making meatballs with my father, and on the other hand, theorizing eating animals as an exercise of cruelty and power, as exemplified by the sorrow I felt as a teenager when I confronted the reality that I was consuming animals whom I used to call my friends.

These theoretical cracks and ruptures raise critical questions about the convergences of theory and pedagogical practice in critical food systems education (CFSE) and animal-focused education (AFE). Questions about animal lived experience are notably absent in much critical food studies and critical food systems education literature, which tends to focus explicitly on cultural, as well as social axes of difference (race, class, gender, disability) without consideration of animal oppression, even when critical approaches are embraced (e.g., Sachs & Patel-Campillo, 2014; Sumner, 2016; Williams-Forson & Wilkerson, 2011). In contrast, some work from an animal advocacy or liberationist perspective does not significantly engage with questions about the cultural significance of eating animals or the social justice dimensions of animal agriculture, both of which raise important questions about the intersections of animality, species, race, class, and culture (e.g., Harper, 2010, 2012; Thirukkumaran, 2017; Wrenn & Johnson, 2013; Wrenn, 2017). The rest of this chapter is thus guided by the following question:

- How do we, as critical food systems and critical animal-focused educators, navigate the tensions between conceptualizing eating animals as a cultural practice and eating animals as an act of power and violence?

Although there are diverse theoretical frameworks that might inform such a pedagogical project, I develop an intersectional cultural humility approach, which embraces self-awareness, openness, and transcendence (Ortega & Faller, 2011) as well as includes critical animal studies (Corman & Vandrovcová, 2014), ecofeminist (Adams, 1990/2015; Gruen & Jones, 2015), humane education (Arbour, Signal, & Taylor, 2009; Humes, 2008; Oakley, 2019), and race-conscious vegan perspectives (Feliz Brueck, 2017; Harper, 2010; Ko & Ko, 2017; Polish, 2016). This approach is not meant to be prescriptive, but rather to explore possibilities for a critical food systems education that recognizes the cultural significance of meat eating for diverse groups of people, acknowledges that racialized and low-income people are marginalized in food systems and do not always have access to vegan/vegetarian diets, and yet also challenges the status quo of carnism (Joy, 2010) To conclude, I draw out the implications of this research for the field of EE, particularly for scholars working in animal-focused and/or critical food systems education. The ultimate goal is to explore the possibilities of a more liberated human-animal politic.

Positionality

I come to this work as a White, middle-class, cisgender woman, who is also a member of the Spanish diaspora in California. For over 25 years, I had not consumed red meat and was vegetarian most of my adult life, adopting veganism until I recently experienced health issues. In 2018, I therefore had to incorporate some flesh and eggs into my diet to care for my own body. I still eschew dairy products for both ethical and health reasons. I initially chose to be vegetarian because of my love for animals, but in the last few years also became significantly more attuned to the social justice issues connected to industrialized animal agriculture, as well as large-scale monoculture crop production. My positionality is critically significant here, as my own cultural food practices have shaped my thinking, and thus my pedagogy related to teaching about the role of animals in industrialized food systems. I am also an educator in southern California's ethnically, racially, and socioeconomically diverse Inland Valley. There, women of color primarily populate my courses and many of them are immigrants or first-generation born in the US. They thus bring diverse cultural perspectives and practices to my classroom that I seek to both honor and challenge since the consumerist and individualistic neoliberal logics that dominate contemporary society also influence my students (also see Lloro-Bidart, 2018a, 2018b; Lloro-Bidart, 2019). I thus work to "...heed a foundational concept in critical pedagogy that educators need to start where students *are*, not where we might wish them to be" (Russell, 2019, p. 38 italics in original). This approach taps into students' prior knowledge, as well as provides multiple entry points for challenging anthropocentrism (Russell, 2019).

Interspecies Food Justice: A Nascent Intersectional Cultural Humility Framework

Intersectionality

Intersectionality is a complex theory and methodology with roots in Black feminism (Collins, 1990; Combahee River Collective, 1977/1993; Crenshaw, 1989, 1991) and ecofeminism (Adams, 1990/2015; Adams & Donovan, 1995; Adams &

Gruen, 2014). In this chapter, the two most salient aspects of intersectionality are: (1) the recognition in Black feminist theory that single-axis approaches are reductive and do not take into consideration the many intersecting categories of difference that shape social identities and life and (2) the acknowledgement in ecofeminism that "the ideology which authorizes oppressions based on gender (class, race, ethnicity, age, etc.) is the same ideology which sanctions the oppression of animals and nature" (Hovorka, 2015, p. 5; also see Lloro-Bidart & Finewood, 2018 and Maina-Okori, Koushik, & Wilson, 2018 for recent overviews). Thus, the intersectional approach I take in this chapter considers the salience of social categories of difference when teaching about food and animals, as well as recognizes that people, animals, and the environment are marginalized in food consumption and production systems (e.g., Oakley, 2019; Russell, 2019). This approach follows the work of intersectional vegan Black feminists insofar as it does not assume that the marginalization of people, animals, and the environment is equal, but rather examines the ways in which the same oppressive power structures differentially impact each of these groups (e.g., Harper, 2010, 2012; Ko & Ko, 2017; Thirukkumaran, 2017).

Cultural Humility

Although "cultural humility" first developed in fields like social work, health care, and child welfare studies in response to critiques of reductive "cultural competence frameworks" (Abrams & Moio, 2009; Furlong & Wight, 2011; Tervalon & Murray-Garcia, 1998), it can be applied to critical food systems pedagogies and praxis. According to Ortega and Faller,

> A cultural humility approach advocates for incorporating multicultural and intersectional understanding and analyses to improve practice, since together these concepts draw attention to the diversity of the whole person, to power differences in relationships…to different past and present life experiences… (Ortega & Faller, 2011, p. 32).

The authors also go on to identify three specific elements of a cultural humility for social work practitioners, which can be applied to educational practice: (1) self-awareness, which acknowledges how people's cultural perspectives shape "the lenses through which they view the world" (Ortega & Faller, 2011, p. 32); (2) openness, which focuses on the fact that people can never know all that there is to know about the world and that we may consciously or unconsciously be influenced by aspects of the world outside of our awareness; and (3) transcendence, which considers how no matter how much we attempt to know the world or other people, they are far more dynamic than we can imagine. In critical food systems education, an intersectional cultural humility approach has the potential to influence "practice," which not only includes pedagogical practice in terms of the curriculum and teaching methods educators select, but also the implicit and explicit ways we interact with our students when we teach about food systems.

Intersectional Cultural Humility and the Significance of Other Animals

While Ortega and Faller (2011) mostly embrace anthropocentric understandings of intersectionality, multiculturalism, and cultural humility, their approach can be expanded to include other animals and the environment. First, educators can critically consider their own positionalities in relation to food systems and how they impact their own and others' teaching practice. This kind of self-awareness is not only valuable in the sense that it helps reflect on how educators see the world, but it also sets a critical example for students, who also need to develop self-awareness with respect to their own positioning in society, including within food systems. Second, when educators practice openness and provide their students with the requisite skills to practice openness, they can expand students' budding critical perspectives to include marginalized groups of humans, other animals, and the environment. In the case of an educator who has already herself embraced intersectional veganism, a willingness to hear student voices, however much those voices might marginalize animals, is also critical to fostering openness (e.g., Oakley, 2019). This is especially true in the multicultural and socioeconomically diverse classroom where students bring differing cultural and class-based perspectives to bear on the curriculum (e.g., Stapleton, 2015). And finally, transcendence in critical food systems education is key because it provides a space to acknowledge that we cannot ever fully know other people or their worlds. In the following sections of this chapter, I thus explore these aspects of an intersectional cultural humility approach in more detail, providing concrete examples of each by discussing my pedagogy and research on the course in relation to the research literature.

Self-Awareness

Although Harper (2010, 2012) and Kim (2015) do not specifically engage a cultural humility framework in their scholarship that closely examines the interconnectedness of race and species, they have deeply challenged me to think more critically about my own positionality when I teach about industrialized food systems. Like many other critical food justice scholars (e.g., Agyeman & McEntee, 2014; Alkon & Norgaard, 2009; Sachs & Patel-Campillo, 2014; Stapleton, 2015), Harper (2010) emphasizes that space is mediated by ability, class, gender, and race, which in turn mediate access to food. Thus, social axes of difference affect "individual and place identities, including one's philosophy of what counts as a moral food system" (Harper, 2010, p. 155). Given Harper's (2010, 2012) intersectional Black vegan perspective, she is also concerned with animal rights and wellbeing, unlike some other critical feminist and/or critical race scholars of food. Not only do students have ideas about what counts as a moral food system, but as educators, we do, too. It is critical, therefore, that we continually evaluate our own positionalities in relation to food systems, as well as encourage our students to do the same.

Kim's (2015) ethics of mutual avowal, which embraces a multi-optic vision and challenges single-optic visions that foreground zero-sum thinking, also offers important insights into self-awareness. She explains that multi-optic practices challenge single optic ones by subverting the belief that "if animals 'win' then people must 'lose'" and by "…encourage[ing] a reorientation toward an *ethics of mutual avowal*, or open and active acknowledgement of connection with other struggles ('This matters to me and relates to me' instead of 'That has nothing to do with me.')" (Kim, 2015, p. 20). When simultaneously addressing issues of animal rights and multiculturalism, both of which unfold in debates about food (e.g., the Makah whale hunt, the East Asian dog meat trade, East Asian uses of shark fins in food), Kim's ethics of mutual avowal can challenge educators to embrace positionalities that connect us to the struggles of myriad others, as well as to develop pedagogies to help our students do the same. In my own classroom, for example, I often share with my students my childhood experiences with sheep (and other animals), as well as my journey to vegetarianism and veganism to provide a platform for them to feel comfortable doing the same. In some cases, they also discuss helping or observing their parents slaughtering animals, offering a unique cultural perspective on the use of animals for food. Some even describe close, intimate relationships they have had with food animals or share their own journeys to vegetarianism/veganism. As I have written elsewhere, discussions centering student experiences provide rich points of discussion,

> One Filipino student noted, for example, that wasting animal parts at the point of slaughter/production or throwing meat in the trash is disrespectful in his culture. Besides the obvious ethical questions he raised about meat eating, he also pointed to another salient topic: food waste is a significant issue in the US particularly food waste that contains animal products because of the amount of energy, resources, and suffering that go into producing food containing animal products. Further, this discussion fostered openness as students realized that the other students in the class did not necessarily hold the same perceptions of meat eating as they did (Lloro-Bidart, 2019, pp. 62–63).

Although the student depicted above did not embrace a vegan abolitionist perspective aimed at ending all human and animal suffering in food systems, his unique cultural perspectives in some ways subvert neoliberal ideologies of over-production, consumption, and waste that wreak havoc on multiple living entities and the environment, as well as challenge dominant narratives regarding which parts of animals it is acceptable to consume or waste (e.g., Alkon and Guthman (2017) and Kim (2015) on resisting neoliberal elites).

Openness

In a traditional cultural humility approach, openness acknowledges that people can never know all that there is to know about the world and that we may consciously or unconsciously be influenced by aspects of the world outside of our awareness. Expanding this concept to include other animals and the environment includes knowing that we can never fully understand the other (human, nonhuman, or otherwise), as well as recognizing that myriad living and nonliving entities influence us, though they may remain outside our awareness. In the context of critical food systems pedagogy, this means that educators can remain attuned to and grapple with the ways in which their own knowledge of the world, including knowledge of food systems, is limited. Further, it opens up a space to explore how our own ideas about what constitutes an ethical, just, and moral food system are shaped by aspects of the world that are both within and outside of our awareness (e.g., Bailey, 2007; Harper, 2012).

For example, I spent 4 years teaching critical food studies while being vegetarian/vegan. During that time, I regularly struggled with how to teach about animal oppression in food systems without alienating my students, whom I knew were mostly omnivores based on brief class polls designed to understand what kinds of identities they embraced in relation to meat. I also feared that I could easily alienate them because empirical research has demonstrated high student resistance to vegetarianism (e.g., Darst & Dawson, 2019) and because I know eating meat is significant in my own culture and assumed it would be the same for my students. I had very limited food options when I stayed with my father's side of the family in a rural village in Northern Spain, for example, simply because nearly everything is bathed in animal fat or animal parts and meat is the center of the multicourse meal. Whenever my aunt cooked a special meal for me—*judias* (beans) or *tortilla espanola* (potato omelette)—I felt pangs of guilt in my chest for rejecting my family's food. Further, I have encountered countless family members here in the US who were initially ostracized by my vegetarianism/veganism just because they observed I was abstaining from meat eating. Although I never initiate conversations about animal oppression when gathering for a meal, the simple refusal to eat meat makes many people uncomfortable (e.g., Rothberger, 2014).

And yet, reflecting on my pedagogy, I was also very unsettled by the thought of omitting a discussion of animal oppression in food systems from my course. Therefore, after ongoing reflection, I came to the realization that people continue to eat meat for many complex reasons, even when they are disturbed by and want to do something about animal oppression. Indeed, my own recent return to meat eating indelibly illustrates these complexities. As Rice (2013) highlights, much of this is due to willful ignorance and incontinence. Willful ignorance involves partial knowing about where meat comes from, but a turning away from or denial of the truth, whereas incontinence refers to knowledge of the truth and continual meat eating with a sense of discomfort. Incontinent meat eaters typically feel bad about their dietary habits and might refer to meat eating as "a guilty pleasure" (Rice, 2013; also see Coetzee, 1999).

Although Rice's (2013) model is critically significant, it does not necessarily take into account the salience of ability, class, culture, gender, race, and other social categories of difference and how they shape our identities in relation to food, as well as our ability to access food (Bailey, 2007; Lloro-Bidart, 2019; Stapleton, 2015). An educator not practicing openness might read her students as ignorant, willfully ignorant, or incontinent if they refuse to consider the benefits of a vegetarian/vegan diet. And in some cases, this might be a correct assumption. In contrast, an educator embracing openness would recognize the complexities of her students' lived experience and realize that she cannot fully know the totality of these experiences, including how they shape students' understandings of and relationships with food systems. Many of my students, for example, hail from working-class backgrounds. Even if they wanted to shop at natural food markets to purchase meat and dairy substitutes, many of these products are outside of their financial reach (e.g., Harper, 2012). Middle- and upper-class Whites might suggest that beans and rice are cheap, high quality vegan foods that anyone can affordably access; however, this kind of approach negates the histories of the poor and people of color who were historically denied access to meat and often had to eat beans and rice out of necessity. Further, students are often concerned about family and social isolation if they adopt a vegetarian or vegan diet. For them, meat is culturally significant and abstaining from it might stress important relationships with humans, as I described above (e.g., Lloro-Bidart & Sidwell, 2019; Rothberger, 2014; Stapleton, 2015). One of my students who had embraced vegetarianism, for example, wanted to get her parents on board so that they could "share the same food" and another expressed that it would be impossible to be vegetarian because her older son would never go for it (Lloro-Bidart, 2019). And finally, for some people dietary restrictions (e.g., gluten, eggs, dairy, soy, and nuts) and/or serious health conditions require they consume meat in order to avoid, mitigate, or remediate illness. This, of course, does not mean that the educator should omit a discussion of animal exploitation and oppression in her courses. Rather, it suggests that a multi-axis approach would take into account the complexities of meat eating, particularly in industrialized systems. As Bailey notes, "The goal must be, then, to balance an appreciation for how eating practices are tied to identity, with a critique of it" (Bailey, 2007, p. 57).

Transcendence

Transcendence refers to the dynamism of the world and other people and involves an acknowledgment that we can never fully know them. While in graduate school, my fellow classmates working in K-12 schools would often share frustrating experiences, many of them working hard to connect this experience to critical readings and theories. However, sometimes my classmates invoked rather reductive explanations, implying that they understood more about their students and students' families than they actually did. In many ways, this reaction is completely understandable given how little support K-12 educators receive in the US. Effectively meeting students where they were, my doctoral advisor would always pose the following question to encourage critical and reflective thinking in classrooms, schools, and beyond, "Do you *know* that to be true?" A piece of the critical mindset she hoped to foster included recognizing that it is often impossible to fully know other people's experiences.

Since that time, I have found my advisor's question to be useful when teaching about food systems. I ask students, for example, "What do we actually *know* to be true about people who use food assistance?" (versus the myths perpetuated in the media) or, "What do we actually know about the lives of animals and workers in CAFOs?" To address the first question, my students engage in an activity where they assume the identity of a fictitious family living near the federal poverty line in the US. I provide them the family size and income, but they get to decide their family's ethnic/racial identity, whether or not they have transportation, and if all or some of their family members are vegetarian/vegan. They then have to use the US's Supplemental Nutrition Assistance Program guidelines to calculate whether or not their family qualifies for food assistance. Students create a grocery shopping list based on this budget, making attempts to select both culturally appropriate food and food that might reduce or eliminate animal and human exploitation. This activity encourages transcendence because it provides an opportunity for students to recognize that they may not fully understand the lives and experiences of people living in poverty, including the challenges they might face in adopting a vegetarian, vegan, or any other kind of special diet. To explore the question about CAFOs, my students engage with readings and videos that seek to illuminate people's and animal's experiences inside these industrialized and mechanized facilities, since it is difficult or impossible to actually access them in real-life. Carefully selected classroom materials and a visit to a local sustainable farm that houses farm animals, some of whom are later slaughtered and sold at the farm's store, provide a window into alternative systems, as well as the inherent flaws in any of these systems. In this assignment, my students are encouraged to think about how transcendence can be applied not only to different groups of people, but also to other invisibilized species. Further, they are challenged to think about how their own cultural practices might contribute to human and animal suffering (see Lloro-Bidart, 2018a; Lloro-Bidart, 2019).

Conclusion

To conclude, this chapter has begun to develop an intersectional cultural humility approach to critical food systems education that includes other animals. By embracing self-awareness, openness, and transcendence, educators can acknowledge and teach about animal oppression in food systems, while also considering how culture, race, class, and other social categories of difference influence and shape meat eating, especially in carnist societies (Joy, 2010). These kinds of multi-optic approaches, which resonate with ecofeminism (Adams, 1990/2015; Adams & Gruen, 2014) and intersectionality (Collins, 1990; Crenshaw, 1989, 1991), do not subvert critique by accepting animal oppression (e.g., meat eating) as simply an important (and therefore unquestionable) cultural practice, nor do they accept vegan/vegetarianism as "cruelty-free" and accessible to everyone. Rather, they hope to foster critical, reflective, and reflexive dialogues that work to discourage zero-sum thinking and foster solidarity with others. Some vegan educators operating from an animal liberationist perspective might argue that in order to achieve solidarity with the animal other, we must teach our students to abstain from consuming all animal products. However, like Russell (2019), I advocate that starting with where students are is critical. My students' lives are significantly shaped by their class position, race, gender, and in some cases ability status and size. Some of them simply cannot afford a vegetarian/vegan diet, while others might face being ostracized by family members (and their culture) if they give up consuming animal products. Yet this does not mean that my students do not care about animal oppression or that they do not critique systems of oppression.

References

Abrams, L. S., & Moio, J. A. (2009). Food studies race theory and the cultural competence dilemma in social work education. *Journal of Social Work Education, 45*(2), 245–259.

Ackerman, Z., Musil, R. K., McAuliffe, E., Brunson, E., & Reynolds, L. (2017). *Fowl matters: Public health, environmental justice, and civil action around the broiler chicken industry*. Bethesda, MD: Rachel Carson Council.

Adams, C. J. (1990/2015). *Sexual politics of meat: A feminist-vegetarian critical theory*. New York, NY: Continuum Press.

Adams, C. J., & Donovan, J. (1995). *Animals and women: Feminist theoretical explorations*. Durham, NC: Duke University Press.

Adams, C. J., & Gruen, L. (2014). *Ecofeminism: Feminist intersections with other animals & the earth*. New York, NY: Bloomsbury.

Agyeman, J., & McEntee, J. (2014). Moving the field of food justice forward through the lens of urban political ecology. *Geography Compass, 8*(3), 211–220.

Alkon, A. H., & Guthman, J. (2017). *The new food activism: Opposition, cooperation, and collective action*. Oakland, CA: University of California Press.

Alkon, A. H., & Norgaard, K. M. (2009). Breaking the food chains: An investigation of food justice activism. *Sociological Inquiry, 79*(3), 289–305.

Arbour, R., Signal, T., & Taylor, N. (2009). Teaching kindness: The promise of humane education. *Society & Animals, 17*, 136–148.

Bailey, C. (2007). We are what we eat: Feminist vegetarianism and the reproduction of racial identity. *Hypatia, 22*(2), 39–59.

Blecha, J., & Leitner, H. (2014). Reimagining the food systems, the economy, and urban life: New urban chicken-keepers in US cities. *Urban Geography, 35*(1), 86–108.

Coetzee, J. M. (1999). *The lives of animals*. Princeton, NJ: Princeton University Press.

Collins, P. H. (1990). *Black feminist thought: Knowledge, consciousness, and the politics of empowerment*. New York: Routledge.

Combahee River Collective. (1977/1993). A black feminist statement. In G. Hill, P. B. Scott, & B. Smith (Eds.), *All the women are white, all the blacks are men, but some of us are brave: Black women's studies* (pp. 13–22). New York, NY: Feminist Press at CUNY.

Corman, L. (2017). Ideological monkey wrenching: Nonhuman animal politics beyond suffering. In D. Nibert (Ed.), *Animal oppression and capitalism, volume 2: The oppressive and destructive role of capitalism* (pp. 252–269). Santa Barbara, CA: Praeger Press.

Corman, L., & Vandrovcová, T. (2014). Radical humility: Toward a more holistic critical animal studies pedagogy. In A. Nocella II, J. Sorenson, K. Socha, & A. Matsuoka (Eds.), *Defining critical animal studies: An introduction to an intersectional social justice approach to animal liberation* (pp. 135–157). New York, NY: Peter Lang.

Crenshaw, K. W. (1989). *Demarginalizing the intersection of race and sex: A black feminist critique of antidiscrimination doctrine, feminist theory, and antiracist politics* (pp. 139–167). Chicago, IL: University of Chicago Legal Forum.

Crenshaw, K. W. (1991). Mapping the margins: Intersectionality, identity politics, and violence against women of color. *Stanford Law Review, 43*(6), 1241–1299.

Darst, R. G., & Dawson, J. I. (2019). Putting meat on the (classroom) table: Problems of denial and communication. In T. Lloro-Bidart & V. S. Banschbach (Eds.), *Animals in environmental education: Interdisciplinary approaches to curriculum and pedagogy* (pp. 215–236). London: Palgrave Macmillan.

Feliz Brueck, J. (2017). *Veganism in an oppressive world: A vegans of color community project.* Cheyenne: Sanctuary Publishers.

Food Empowerment Project. (2018). Factory farm workers. Retrieved from http://www.foodispower.org/factory-farm-workers/

Furlong, M., & Wight, J. (2011). Promoting "critical awareness" and critiquing "cultural competence": Towards disrupting received professional knowledge. *Australian Social Work, 64*(1), 38–54.

Gaard, G. (2001). Tools for a cross-cultural feminist ethics: Exploring ethical contexts and contents in the Makah whale hunt. *Hypatia, 16*(1), 1–26.

Gruen, L., & Jones, R. C. (2015). Veganism as aspiration. In B. Bramble & B. Fischer (Eds.), *The moral complexities of eating meat* (pp. 153–171). Oxford: Oxford University Press.

Harper, A. B. (2010). Race as a "feeble matter" in veganism: Interrogating whiteness, geopolitical privilege, and consumption philosophy of cruelty-free products. *Journal of Critical Animal Studies, 8*(3), 5–27.

Harper, A. B. (2012). Going beyond the normative white "post-racial" vegan epistemology. In P. Williams-Forson & C. Counihan (Eds.), *Taking food public: Redefining foodways in a changing world* (pp. 155–174). New York, NY: Routledge.

Hovorka, A. (2015). The *Gender, Place, and Culture* Jan monk distinguished annual lecture: Feminism and animals: Exploring interspecies relations through intersectionality, performativity, and standpoint. *Gender, Place, & Culture, 22*(1), 1–19.

Humes, B. (2008). Moving toward a liberatory pedagogy for all species: Mapping the need for dialogue between humane and anti-oppressive education. *Green Theory & Praxis: The Journal of Ecopedagogy, 4*(1), 65–85.

Inge, L. (2014). *Female butchers are slicing through the meat world's glass ceiling.* Washington, DC: National Public Radio. Retrieved from https://www.npr.org/sections/thesalt/2014/12/07/366642071/women-butchers-are-slicing-through-the-meat-worlds-glass-ceiling

Joy, M. (2010). *Why we love dogs, eat pigs, and wear cows.* San Francisco, CA: Carnari Press.

Kim, C. J. (2015). *Dangerous crossings: Race, species, and nature in a multicultural age.* Cambridge, MA: Cambridge University Press.

Ko, A., & Ko, S. (2017). *Aphro-ism: Essays on pop culture, feminism, and black veganism from two sisters.* New York, NY: Lantern Books.

Lloro-Bidart, T. (2015). "Culture *as* ability": Organizing enabling educative spaces for humans *and* animals. *Canadian Journal of Environmental Education, 20*, 93–108.

Lloro-Bidart, T. (2018a). A feminist posthumanist ecopedagogy in/for/with animalscapes. *Journal of Environmental Education, 49*(2), 152–163.

Lloro-Bidart, T. (2018b). Cultivating affects: A feminist posthumanist analysis of invertebrate and human performativity in an urban community garden. *Emotion, Space, and Society, 27*, 23–30.

Lloro-Bidart, T. (2019). Intersectional & interdisciplinary approaches to interspecies food justice pedagogies. In T. Lloro-Bidart & V. Banschbach (Eds.), *Animals in environmental education: Interdisciplinary approaches to curriculum and pedagogy* (pp. 53–76). New York: Palgrave Macmillan.

Lloro-Bidart, T., & Finewood, M. (2018). Looking outward and inward: What feminist theory offers the environmental studies and sciences. *Journal of Environmental Studies and Sciences, 8*(2), 141–152.

Lloro-Bidart, T., & Semenko, K. (2017). Toward a feminist ethic of self-care for environmental educators. *Journal of Environmental Education, 48*(1), 18–25.

Lloro-Bidart, T., & Sidwell, C. M. (2019). Relational identity and intergenerational learning in an undergraduate critical food studies course. *Journal of Environmental Education, 51*, 200. https://doi.org/10.1080/00958964.2019.1625298

Lupinacci, J., & Happel-Parkins, A. (2018). Food for a common(s) curriculum: Learning to recognize and resist food enclosures. In S. Rice & A. G. Rud (Eds.), *Educational dimensions of school lunch: Critical perspectives* (pp. 91–115). Cham: Palgrave Macmillan.

Maina-Okori, N. M., Koushik, J. R., & Wilson, A. (2018). Reimagining intersectionality in environmental and sustainability education: A critical literature review. *Journal of Environmental Education, 49*(4), 286–296.

Oakley, J. (2019). What can an animal liberation perspective contribute to environmental education? In T. Lloro-Bidart & V. Banschbach (Eds.), *Animals in environmental education: Interdisciplinary approaches to curriculum and pedagogy* (pp. 19–34). New York, NY: Palgrave Macmillan.

Oakley, J., Watson, G. P. L., Russell, C. L., Cutter-Mackenzie, A., Fawcett, L., Kuhl, G., et al. (2010). Animal encounters in environmental education research: Responding to the question of the animal. *Canadian Journal of Environmental Education, 15*, 86–102.

Ortega, R. M., & Faller, K. C. (2011). Training child welfare workers from an intersectional cultural humility perspective. *Child Welfare, 90*(5), 27.

Pachirat, T. (2011). *Every twelve seconds*. New Haven, CT: Yale University Press.

Pedersen, H., & Stănescu, V. (2012). Series editors' introduction: What is "critical" about Animal studies? From the animal 'question' to the animal 'condition'. In K. Socha (Ed.), *Women, destruction, and the avant-garde: A paradigm for animal liberation* (pp. ix–xi). Amsterdam: Rodopi.

Polish, J. (2016). Decolonizing veganism: On resisting whiteness and racism. In J. Castricano & R. Simonsen (Eds.), *Critical perspectives on veganism* (pp. 373–391). New York, NY: Palgrave Macmillan.

Pollan, M. (2002, November 10). An animal's place. *New York Times Magazine*. Retrieved from https://michaelpollan.com/articles-archive/an-animals-place/

Pollan, M. (2006). *The omnivore's dilemma*. New York, NY: Bloomsbury.

Potts, A. (2016). *Meat culture*. Boston, MA: Brill.

Rice, S. (2013). Three educational problems: The case of eating animals. *Journal of Thought, 48*(2), 112–127.

Rothberger, H. (2014). A comparison of attitudes toward meat and animals among strict and semi-vegetarians. *Appetite, 72*, 98–105.

Rowe, B. (2013). It IS about chicken: Chick-fil-a, intersectionality, and gastro-aesthetic pedagogy. *Journal of Thought, 48*(2), 89–111.

Rudy, K. (2012). Locavores, feminism, and the question of meat. *The Journal of American Culture, 35*(1), 26–36.

Russell, C. (2005). 'Whoever does not write is written': The role of 'nature' in post-post approaches to environmental education research. *Environmental Education Research, 11*(4), 433–443.

Russell, C. (2019). An intersectional approach to teaching and learning about humans and other animals in educational contexts. In T. Lloro-Bidart & V. Banschbach (Eds.), *Animals in environmental education: Interdisciplinary approaches to curriculum and pedagogy* (pp. 35–52). New York, NY: Palgrave Macmillan.

Russell, C., & Fawcett, L. (2013). Moving margins in environmental education research. In R. B. Stevenson, M. Brody, J. Dillon, & A. E. J. Wals (Eds.), *The international handbook for research on environmental education* (pp. 369–374). New York, NY: Routledge.

Russell, C., & Semenko, K. (2016). We take "cow" as a compliment: Fattening humane, environmental, and social justice education. In E. Cameron & C. Russell (Eds.), *The fat pedagogy reader: Challenging weight-based oppression through critical education* (pp. 211–220). New York, NY: Peter Lang.

Sachs, C., & Patel-Campillo, A. (2014). Feminist food justice: Crafting a new vision. *Feminist Studies, 40*(2), 396–410.

Spannring, R. (2017). Animals in environmental education research. *Environmental Education Research, 23*(1), 63–74.

Spannring, R., & Grušovnik, T. (2018). Leaving the meatrix? Transformative learning and denialism in the case of meat consumption. *Environmental Education Research, 25*, 1190. https://doi.org/10.1080/13504622.2018.1455076

Stapleton, S. R. (2015). Food, identity, and environmental education. *Canadian Journal of Environmental Education, 20*, 12–24.

Sumner, J. (2016). *Learning, food, & sustainability: Sites for resistance and change.* New York, NY: Palgrave Macmillan.

Thirukkumaran, M. R. (2017). The "V" Word: An Inquiry into Vegan Student Experience in Calgarian Schools (Unpublished doctoral thesis). University of Calgary, Calgary, AB. doi:10.11575/PRISM/28421.

Tervalon, M., & Murray-Garcia, J. (1998). Cultural humility versus cultural competence: A critical distinction in defining physician training outcomes in multicultural education. *Journal of Health Care for the Poor and Underserved, 9*(2), 117–125.

Werft, M. (2015, November 25). Why it's worth it to meet your meat this Thanksgiving. *Global Citizen.* Retrieved from https://www.globalcitizen.org/en/content/why-its-worth-it-to-meet-your-meat-this-thanksgivi/

Williams-Forson, P., & Wilkerson, P. (2011). Intersectionality and food studies. *Food, Culture & Society: An International Journal of Multidisciplinary Research, 14*(1), 7–28.

Wrenn, C. L. (2017). Trump veganism: A political survey of American vegans in the era of identity politics. *Societies, 7*(4), 1–13.

Wrenn, C. L., & Johnson, R. (2013). A critique of single-issue campaigning and the importance of abolitionist vegan advocacy. *Food, Culture, & Society, 16*(4), 651–668.

Wright-Maley, C. (2011). Meet them at the plate: Reflections on the eating of animals and the role of education therein. *Critical Education, 2*(5), 1–21.

Afterword

Inextinguishable Realities: Educating for Environmental Justice in Climate-Changing Worlds

Donna Houston

On September 20, 2019, as the brutal and devastating bushfires that would burn through the Australian spring and summer flared into life, hundreds of thousands of school children, families, faith organisations and workers marched in climate rallies across the nation. In Sydney, the rally of over 100,000 people took place on an unseasonably hot Friday afternoon, prompting a number of politicians in the Federal and New South Wales state governments to accuse teachers of allowing "climate politics" to infiltrate classrooms. The Federal Minster for Education went as far as to link the striking students with declining national test scores.[1] School children refuted declarations by politicians that they should be in class, arguing that they were putting their school lessons into responsible civic action. They creatively illustrated their fears and frustrations with a sea of hand-made signs. A popular placard echoed the words of Swedish teenage climate activist Greta Thunberg who inspired the School Strike 4 Climate Action movement: "Our House is on Fire!" (Fig. 1)

The climate rallies spoke to palatable intergenerational and community frustration with the Federal government's policy inaction on climate change. In the months that followed, when the smoke from the bushfires rolled over Sydney in toxic waves, tens of thousands of protesters wearing face masks gathered again in the city centre to protest the Federal government's downplaying of the climate emergency.[2]

[1] ABC News. (2019). Global climate strikes sees 'Hundreds of Thousands' of Australians Rally Across the Country. Retrieved July 4, 2020, from https://www.abc.net.au/news/2019-09-20/school-strike-for-climate-draws-thousands-to-australian-rallies/11531612.

[2] Stephens, J. (2019). 20000 protest over Sydney's smoke. *Australian Financial Review*. Retrieved July4,2020,fromhttps://www.afr.com/politics/tens-of-thousands-rally-in-sydney-as-fires-burn-20191212-p53j63.

D. Houston
Department of Geography and Planning, Macquarie University, Sydney, NSW, Australia
e-mail: donna.houston@mq.edu.au

A. S. Gkiolmas, C. D. Skordoulis (eds.), *Towards Critical Environmental Education*, Critical Studies of Education 14,
https://doi.org/10.1007/978-3-030-50609-4

Fig. 1 Street art on a wall at Bondi Beach, January 2020. The image depicts the Prime Minister of Australia, Mr Scott Morrison, on a sinking boat with bags of money, amidst rising seas, while the regions around Sydney burn (Photo: Author)

Colloquially known as the "Black Summer", the fires were unprecedented in their scope and severity (Griffiths, 2020). In their report *Summer of Crisis*, the Climate Council of Australia estimated that eighty per cent of Australians were either directly or indirectly impacted by the fires (Climate Council, 2020). More than 11 million hectares of land were burnt, 33 human lives were lost, thousands of homes were destroyed and over a billion animals were killed, not including reptiles, frogs, invertebrates and those animals whom survived the fires only to face thirst, predation and starvation. The Gospers Mountain fire that burned 500,000 hectares just north of Sydney was the largest forest fire ever recorded in Australia (Climate Council, 2020, p. II).

It is difficult to find the words to describe the grief, the loss to the ongoingness of land and life—the repercussions of which will be felt across multispecies generations. For species that have been driven to the edge of extinction in the fiery chaos, these repercussions are final. The late Deborah Bird Rose (n.d. online) whose writings so eloquently spoke to the unthinkable of the Anthropocene, called this form of environmental destruction that outstrips the regenerative capacities of life, "double death":

> So many losses occur that damaged ecosystems are unable to recuperate their diversity. The death of resilience and renewal, at least for a while.

So many extinctions that the process of evolution is unable to keep up. More species die than are coming into being. The death of evolution itself, at least for a while.

The unmaking of country, unraveling the work of generation upon generation of living beings; cascades of death that curtail the future and unmake the living presence of the past.

The death of temporal, fleshy, metabolic relationships across generations and species.

The destruction of the future of one's own death, which starts to collapse along with the future of flourishing others and ecosystems" (ND, online).

For First Nations peoples, the damage wrought on Country by the fires is felt acutely. Lorena Allam, a journalist of Gamilaraay and Yawalaraay descent, poignantly described the damage to the interconnections of culture, history and place on Yuin lands on the south coast of New South Wales:

Some of these places have never burned, not once in my lifetime, let alone all at once. Like you, I've watched in anguish and horror as fire lays waste to precious Yuin land, taking everything with it – lives, homes, animals, trees – but for First Nations people it is also burning up our memories, our sacred places, all the things which make us who we are.

It's a particular grief, to lose forever what connects you to a place in the landscape. Our ancestors felt it, our elders felt it, and now we are feeling it all over again as we watch how the mistreatment and neglect of our land and waters for generations, and the pig-headed foolishness of coal-obsessed climate change denialists turn everything and everyone to ash (2020, online).

It is important to sit with these words because they connect the fires with the ongoing denial of First Nations sovereignty—sovereignty which is expressed through the embodied knowledge of many thousands of generations of living in good relations with the land. Lorena Allam's words resonate with Potawatomi scholar and activist Kyle Powys Whyte's (2017) assertions in his chapter "Our Ancestors' Dystopia Now: Indigenous Conservation and the Anthropocene". Whyte emphasises the particular implications of anthropogenic climate change for Indigenous peoples and communities already bearing the brunt of settler industrial-colonialism on their lands, waters and reciprocal relationships with nonhuman kin:

In the Anthropocene, then, some indigenous peoples already inhabit what our ancestors would have likely characterized as a dystopian future. So we consider the future from what we believe is already a dystopia, as strange as that may sound to some readers. Our conservation and restoration projects are not only about whether to conserve or let go of certain species. Rather, they are about what relationships between humans and certain plants and animals we should focus on in response to the challenges we face, given that we have already lost so many plants and animals that matter to our societies. In this way, indigenous conservation approaches aim at negotiating settler colonialism as a form of human expansion that continues to inflict anthropogenic environmental change on indigenous peoples—most recently under the guise of climate destabilization" (Whyte, 2017, p. 207).

The "Black Summer" fires were eventually quenched with a deluge of rain in early February 2020. What I am recounting in this afterword are some of the inextinguishable realities that remain in its aftermath and in the aftermath of environmental damage more broadly, realities that have important implications for critical environmental education.

As expressed by the striking school students, Deborah Bird Rose, Lorena Allam, Kyle Powys Whyte and indeed by many of the authors in their diverse approaches to pedagogies in/and/for the Anthropocene in this book: we are living amidst unprecedented environmental crises that are planetary in scale, but which locally and unequally impact upon people, places, and nonhumans bound within hierarchies of racial, economic and colonial injustice (Nixon, 2011). Rob Nixon (2011, p. 31) in his book *Slow Violence: Environmentalism of the Poor* writes that in order to respond to these pressing realities, scholars, educators and writers are confronted with "critical choices" about how they can enact "creative bridgework" to engage "with ambitious communicative intent, transnational questions arising from the borderlands between empire, neoliberalism, environmentalism and social justice" (2011, p. 31). For Nixon, the lived realities of global heating, carboniferous capitalism and toxic pollution produce temporal dissonances "between short-lived actions and long-lived consequences" which play out across silent and invisible aftermaths—what he terms as the accumulation of "slow violence" (p. 41). For Nixon, the "critical choices" relate to how we can respectfully compose together socially and environmentally just worlds in response to living in the continuing aftermath of slow violence.

Environmental justice education involves making visible the toxic impacts, mass biodiversity losses, human and nonhuman suffering and climate instability of the Anthropocene. It involves making connections across science, social science and humanities curricula so that environmental justice forms the "creative bridgework" of forming new synergies, alliances and connections between historically and culturally diverse more-than-human communities. It involves sustaining the critique of colonialism, racial capitalism and extractive economics, which are co-producing worsening social inequities and which are further degrading environments on a warming planet. Finally, environmental justice requires holding space for experiences and knowledges that have been historically silenced through race, gender and economic oppression and learning how to become accountable to them. As Kyle Whyte (2017) evocatively argues, where appropriate and when offered, First Nations communities can offer significant guidance. First Nations peoples continue to sustain their sovereign legal orders and more-than-human kin in the long shadow of environmental destruction. Whyte highlights three conservation projects in the Great Lakes Region—to restore *nmé* (Great Lakes Sturgeon), *manoomin* (wild rice) and the protection of *nibi* (water) as powerful examples of such reparative work (pp. 210–212).

The Australian fire grounds are showing tentative signs of life: green shoots have sprouted from charred trees and soils, in some places the animals are returning, and there is hopeful news that some species thought to have been entirely perished (such as the Kangaroo Island Dunnart) may have survived. But the scorched rainforests, woodlands, wetlands and coastal heathlands are irrevocably changed and changing. Symptomatic of what fire historian Stephen Pyne (2020) calls the *Pyrocene*, the increasing scale and intensity of the fires and the decreasing intervals between them are fundamentally altering the structure and multispecies interrelations of entire ecosystems. The bushfires generated their own weather and are likely to have released 830 million tonnes of carbon dioxide into the atmosphere, far more than

Australia typically emits in a year.[3] It is perhaps of little surprise that white Australia has taken a sudden interest in the cultural fire practices of First Nations peoples. Cultural burning, Shaun Hooper (Wiradjuri) writes, is "one of the ways Aboriginal people maintain their relationship with Country". It is different from settler practices of hazard reduction burning because fire is part of reciprocal relations between people, plants and animals:

> This relationship-based approach allows for the involvement of other than human beings such as bettongs, bandicoots, lyrebirds, wombats and brush turkeys who all assist with Cultural burning by turning over and reducing the leaf litter (Hooper, 2020, online).

The revitalisation of cultural burning by First Nations people in different parts of Australia emphasises the importance of decolonial work for respectfully negotiating and enacting environmental justice in the Anthropocene. There are important lessons here for how First Nations people can be materially supported to practice their sovereign knowledges for the benefit of everyone (Hooper, 2020). And there are lessons here too for the kind of interconnected and regenerative pedagogy that is required both in and outside of classrooms. This is a difficult proposition for settler and capitalist societies such as Australia, as Indigenous author Tony Birch writes:

> The challenge to white Australia is to address a question. How does the nation move from a state of colonial anxiety that refuses genuine recognition and engagement to a concept of locating 'Indigenous theories, methodologies and methods at the centre, not the periphery' of our society? While such a shift could ultimately produce 'an ecological philosophy of mutual benefit', getting there will be a real challenge (2017, online).

These are the inextinguishable realities of enacting environmental justice in climate-changing worlds, and they underscore the importance of developing critical environmental pedagogies that work productively with historical and cultural difference.

[3] Morton, A. (2020, April 21). Summer's bushfires released more carbon dioxide than Australia does in a year. *The Guardian*. Retrieved July 5, 2020, from https://www.theguardian.com/australia-news/2020/apr/21/summers-bushfires-released-more-carbon-dioxide-than-australia-does-in-a-year#:~:text=Australia's%20devastating%20bushfire%20season%20is,according%20to%20a%20government%20estimate.&text=It%20is%20estimated%2096%25%20of,absorbed%20in%20regrowth%20by%202019.

References

Allam, L.. (2020). For First Nations People the bushfires bring a particular grief burning what makes us who we are. *The Guardian*. Retrieved July 5, 2020, from https://www.theguardian.com/commentisfree/2020/jan/06/for-first-nations-people-the-bushfires-bring-a-particular-grief-burning-what-makes-us-who-we-are

Birch, T. (2017). *Climate change, recognition and social place-making*. Sydney, NSW: Sydney Review of Books. Retrieved July 5, 2020, from https://sydneyreviewofbooks.com/essay/climate-change-recognition-and-caring-for-country/

Bird Rose, D. (n.d.). Double death. The multispecies salon. Retrieved July 5, 2020, from http://www.multispecies-salon.org/double-death/

Climate Council of Australia. (2020). Summer of crisis report. Retrieved July 5, 2020, from https://www.climatecouncil.org.au/wp-content/uploads/2020/03/Crisis-Summer-Report-200311.pdf

Griffiths, T.. (2020). Season of reckoning. Australian book review. March (No. 419). Retrieved July 5, 2020, from https://www.australianbookreview.com.au/abr-online/current-issue/758-commentary/6243-season-of-reckoning-by-tom-griffiths

Hooper, S.. 2020 Cultural burning is more than just hazard reduction. IndigenousX. Retrieved July 5, 2020, from https://indigenousx.com.au/cultural-burning-is-about-more-than-just-hazard-reduction/

Nixon, R. (2011). *Slow violence and the environmentalism of the poor*. Cambridge, MA: Harvard University Press.

Pyne, S. (2020). Burning bush, burning earth. Stephen J. Pyne commentaries, otherwise unpublished. Retrieved July 5, 2020, from http://www.stephenpyne.com/blog/posts/35501

Whyte, K. P. (2017). Our ancestors dystopia now: Indigenous conservation and the anthropocene. In U. K. Heise, J. Christensen, & M. Neimann (Eds.), *The Routledge companion to the environmental humanities*. London: Routledge.

Index

A. S. Gkiolmas, C. D. Skordoulis (eds.), *Towards Critical Environmental Education*, Critical Studies of Education 14,
https://doi.org/10.1007/978-3-030-50609-4

The manufacturer's authorised representative in the EU is Springer Nature Customer Service Centre GmbH, Europaplatz 3, 69115 Heidelberg, Germany. If you have any concerns regarding our products, please contact ProductSafety@springernature.com

Printed and bound by CPI Group (UK) Ltd, Croydon, CR0 4YY
30/09/2026
02225465-0019